ACPL ITEM
DISCARDED

621
Lenk
Lenk's laser handbook

LENK'S LASER HANDBOOK

Other McGraw-Hill Reference Books of Interest

Books by John D. Lenk
LENK'S VIDEO HANDBOOK
LENK'S AUDIO HANDBOOK

Handbooks

Benson • AUDIO ENGINEERING HANDBOOK
Benson • TELEVISION ENGINEERING HANDBOOK
Benson and Whitaker • TELEVISION AND AUDIO HANDBOOK
Coombs • PRINTED CIRCUITS HANDBOOK
Croft and Summers • AMERICAN ELECTRICIANS' HANDBOOK
Di Giacomo • DIGITAL BUS HANDBOOK
Fink and Beaty • STANDARD HANDBOOK FOR ELECTRICAL ENGINEERS
Fink and Christiansen • ELECTRONIC ENGINEERS' HANDBOOK
Hicks • STANDARD HANDBOOK OF ENGINEERING CALCULATIONS
Inglis • ELECTRONIC COMMUNICATIONS HANDBOOK
Kaufman and Seidman • HANDBOOK OF ELECTRONICS CALCULATIONS
Stout • MICROPROCESSOR APPLICATIONS HANDBOOK
Stout and Kaufman • HANDBOOK OF MICROCIRCUIT DESIGN AND APPLICATION
Stout and Kaufman • HANDBOOK OF OPERATIONAL AMPLIFIER CIRCUIT DESIGN
Tuma • ENGINEERING MATHEMATICS HANDBOOK
Williams • DESIGNER'S HANDBOOK OF INTEGRATED CIRCUITS
Williams and Taylor • ELECTRONIC FILTER DESIGN HANDBOOK

Other

Bartlett • CABLE TELEVISION TECHNOLOGY AND OPERATIONS
Luther • DIGITAL VIDEO IN THE PC ENVIRONMENT
Mee and Daniel • MAGNETIC RECORDING, VOLUMES I-III
Philips • COMPACT DISC INTERACTIVE

LENK'S LASER HANDBOOK

Featuring
CD, CDV, and CD-ROM
Technology

John D. Lenk

McGRAW-HILL, INC.
New York   St. Louis   San Francisco   Auckland   Bogotá
Caracas   Lisbon   London   Madrid   Mexico   Milan
Montreal   New Delhi   Paris   San Juan   São Paulo
Singapore   Sydney   Tokyo   Toronto

Library of Congress Cataloging-in-Publication Data

Lenk, John D.
  [Laser handbook]
  Lenk's laser handbook : featuring CD, CDV, and CD-ROM technology / John D. Lenk.
    p.   cm. — (Consumer electronic series)
  Includes index.
  ISBN 0-07-037505-4
  1. Compact disc players—Maintenance and repair. 2. Video disc players—Maintenance and repair. I. Title. II. Title: Laser handbook.
  TK7881.75.L47 1992   91-13991
  621.389'32—dc20   CIP

Allen County Public Library
Ft. Wayne, Indiana

Copyright © 1992 by McGraw-Hill, Inc. All rights reserved. Printed in the United States of America. Except as permitted under the United States Copyright Act of 1976, no part of this publication may be reproduced or distributed in any form or by any means, or stored in a data base or retrieval system, without the prior written permission of the publisher.

1 2 3 4 5 6 7 8 9 0   DOC/DOC   9 7 6 5 4 3 2 1

ISBN 0-07-037505-4

The sponsoring editor for this book was Daniel A. Gonneau, the editing supervisor was Nancy Young, and the production supervisor was Suzanne W. Babeuf. This book was set in Times Roman. It was composed by McGraw-Hill's Professional Book Group composition unit.

Printed and bound by R. R. Donnelley & Sons Company.

Information contained in this work has been obtained by McGraw-Hill, Inc., from sources believed to be reliable. However, neither McGraw-Hill nor its authors guarantees the accuracy or completeness of any information published herein and neither McGraw-Hill nor its authors shall be responsible for any errors, omissions, or damages arising out of this information. This work is published with the understanding that McGraw-Hill and its authors are supplying information but are not attempting to render engineering or other professional services. If such services are required, the assistance of an appropriate professional should be sought.

Greetings from the Villa Buttercup!
To my wonderful wife Irene, Thank you
for being by my side all these years!
To my lovely family, Karen,
Tom, Brandon, and Justin.
And to our Lambie and Suzzie,
be happy wherever you are!
To my special readers, may good fortune
find your doorways
to good health and happy things.
Thank you for buying my books and
making me a best seller!
This is book number 72.
Abundance!

CONTENTS

Preface ix
Acknowledgments xi

Chapter 1. Introduction to Laser-Based Disc Players   1.1

1.1. The Laser Disc Scene / *1.1*
1.2. Introduction to the CD System / *1.2*
1.3. Introduction to the CDV System / *1.16*

Chapter 2. Encoding, Decoding, and Optical Readout   2.1

2.1. Encoding a CD / *2.1*
2.2. Channel Modulation (Cutting the Master) / *2.4*
2.3. Decoding the CD / *2.8*
2.4. CD and CDV Optical Pickups (Readouts) / *2.12*
2.5. CDV Track Formats / *2.20*

Chapter 3. User Controls, Operating Procedures, and Installation   3.1

3.1. Operational Safety Checks / *3.1*
3.2. Transit or Shipping Restraints / *3.2*
3.3. External Connections / *3.2*
3.4. General Operating and Installation Notes / *3.4*
3.5. Basic CD Player User Controls / *3.10*
3.6. Basic CDV Player User Controls / *3.12*

Chapter 4. Test Equipment, Tools, and Routine Maintenance   4.1

4.1. Safety Precautions during Service / *4.1*
4.2. Test Equipment / *4.7*
4.3. Tools / *4.8*
4.4. Periodic Maintenance and Disc Care / *4.9*

Chapter 5. Typical CD Player and CD-ROM Circuits   5.1

5.1. Relationship of CD Player Circuits / *5.1*
5.2. Mechanical Functions / *5.2*
5.3. Laser Optics and Circuits / *5.5*
5.4. Laser Signal Processing / *5.8*
5.5. Audio Circuits / *5.9*
5.6. Autofocus / *5.11*
5.7. Laser Tracking / *5.14*
5.8. Turntable Motor Circuits / *5.16*
5.9. Antishock Circuits / *5.18*
5.10. Jump Circuits / *5.19*
5.11. Introduction to CD-ROM / *5.21*

Chapter 6. Typical CDV Player Circuits — 6.1

6.1. Relationship of CDV Player Circuits / *6.1*
6.2. Power-Supply Circuits / *6.5*
6.3. Mute Circuit / *6.5*
6.4. Spindle-Motor Start-up / *6.7*
6.5. System Control / *6.9*
6.6. Start-Up and Laser Control / *6.12*
6.7. Servo Systems / *6.15*
6.8. Focus Servo / *6.17*
6.9. Tracking and Slider Servo / *6.21*
6.10. Tilt Servo / *6.26*
6.11. Disc Signal Processing / *6.27*
6.12. Video Processing / *6.31*
6.13. Spindle Servo / *6.36*
6.14. Video Distribution, NR, and OSD / *6.40*
6.15. Digital Memory / *6.43*
6.16. Analog Audio / *6.44*
6.17. Digital Audio, Analog and Digital Select, and Bilingual / *6.46*
6.18. Helium-Neon Laser Circuits / *6.49*

Chapter 7. Mechanical Operation, Adjustment, and Replacement — 7.1

7.1. CD Vertical Front-Load Mechanical Section / *7.2*
7.2. CD Horizontal Front-Load Mechanical Section (Single Load Motor) / *7.15*
7.3. CD Horizontal Front-Load Mechanical Section (Two Motors) / *7.26*
7.4. CDV Mechanical Section / *7.29*

Chapter 8. CD Player Troubleshooting and Adjustment — 8.1

8.1. The Basic CD Troubleshooting Functions / *8.1*
8.2. The CD Troubleshooting Approach / *8.4*
8.3. CD Electrical Adjustments / *8.5*
8.4. Mechanical Troubleshooting Approach / *8.11*
8.5. Laser Troubleshooting Approach / *8.14*
8.6. Signal-Process Troubleshooting Approach / *8.16*
8.7. Audio Troubleshooting Approach / *8.17*
8.8. Autofocus Troubleshooting Approach / *8.18*
8.9. Laser-Tracking Troubleshooting Approach / *8.20*
8.10. Turntable Troubleshooting Approach / *8.22*

Chapter 9. CDV Player Troubleshooting and Adjustment — 9.1

9.1. The Basic CDV Troubleshooting Functions / *9.1*
9.2. The CDV Troubleshooting Approach / *9.2*
9.3. CDV Electrical Adjustments / *9.2*
9.4. CDV Trouble Symptoms Related to Adjustment / *9.35*
9.5. Power-Supply Troubleshooting Approach / *9.37*
9.6. Spindle-Drive Troubleshooting Approach / *9.40*
9.7. System-Control Troubleshooting Approach / *9.41*
9.8. Servo Troubleshooting Approach / *9.43*
9.9. Video and TBC Troubleshooting Approach / *9.44*
9.10. Audio Troubleshooting Approach / *9.45*
9.11. Digital-Memory (Special Effects) Troubleshooting Approach / *9.45*
9.12. Helium-Neon Laser Troubleshooting and Adjustment / *9.46*

Index I.1   (follows Chapter 9)

PREFACE

This book provides a simplified, practical system of troubleshooting and repair for laser-based consumer electronic equipment now in common use. Although the emphasis is on compact-disc (CD) audio players and CDV (CD video) players (also known as Laserdisc and/or Laservision), the basic principles described here can also be used by technicians and field-service engineers working with other laser-based systems such as CD interactive (CD-I) and CD-ROM drives.

It is virtually impossible in one book to cover detailed troubleshooting and repair for all laser-based equipment. Similarly, it is impractical to attempt such coverage, since rapid technical advances soon make such a book's details obsolete.

To overcome the problem, this book concentrates on a basic approach to CD and CDV player service, an approach that can be applied to any CD or CDV player (both those now in use and those to be manufactured in the future). The same approach can also be applied to CD-I and CD-ROM drives and is based on the techniques found in the author's best-selling troubleshooting books.

Chapter 1 is devoted to the basics of CD players, including their relationship to stereo systems, and to CDV players, including their relationship to TV receivers and monitors. With the basics established, the chapter then describes the technical characteristics for the various models of CD and CDV players now in use.

Chapter 2 describes the encoding and decoding processes used for compact discs. The chapter also describes the basic principles of optical readout used in all laser-based CD and CDV systems. An understanding of these processes and principles is most helpful, even for the very practical technician whose main concern is with efficient troubleshooting.

Chapter 3 describes user controls, operating procedures, and installation of laser-based CD and CDV players. Although CD and CDV players are not difficult to install or operate, the basic procedures are quite different from those of a typical LP phonograph or a VCR.

Chapter 4 describes the test equipment and tools needed for CD and CDV player service. The chapter also discusses routine maintenance for CD and CDV players and CD and CDV discs. Special emphasis is placed on safety (including laser safety) and the relationship of features found in present-day test equipment to specific problems in service of laser-based equipment.

Chapter 5 describes the theory of operation for typical laser-based CD player circuits. By studying the circuits found in Chap. 5, the reader should have no difficulty understanding the schematic and block diagrams of similar CD players.

Circuit descriptions are supplemented with partial schematics and block diagrams that show such important areas as signal flow paths, input-output, adjustment controls, test points, and power-source connections (the most important areas for service). The chapter concludes with an explanation of major differences between consumer audio CD players and computer-oriented CD-ROM drives, including line-by-line connections between a CD-ROM and host computer.

Chapter 6 provides coverage of CDV players that is similar to that for CD players in Chap. 5.

Chapter 7 describes operation of the mechanical sections for typical CD and CDV players. Although many players are covered, the chapter concentrates on mechanical operation of the most popular models, including manufacturer-recommended adjustment and replacement procedures. This information is essential since most CD and CDV player faults are the result of failure of (or tampering with) the mechanical section.

Chapter 8 describes troubleshooting and service notes for a cross section of CD players, including electrical adjustments (to complement the mechanical adjustments described in Chap. 7). Using these examples, the reader should be able to relate the procedures to a similar set of adjustment points on most CD players.

Where it is not obvious, the chapter also describes the purpose and results of the procedures, including waveforms measured at various test points. By studying these waveforms, the reader should be able to identify typical signals found in most players, even though the signals may appear at different points for a particular player.

With adjustments well established, the chapter then describes circuit-by-circuit troubleshooting for laser-based CD players. This approach is based on failure or trouble symptoms and represents the combined experience and knowledge of many CD player service specialists and managers.

Chapter 9 provides coverage of CDV player circuits that is similar to that for CD players in Chap. 8.

John D. Lenk

ACKNOWLEDGMENTS

Many professionals have contributed their talents and knowledge to the preparation of this book. I gratefully acknowledge that the tremendous effort needed to make this book such a comprehensive work is impossible for one person and wish to thank all who have contributed, both directly and indirectly.

I wish to give special thanks to the following: Tom Roscoe, Dennis Yuoka, and Terrance Miller of Hitachi; Thomas Lauterback of Quasar; Donald Woolhouse of Sanyo; John Taylor and Matthew Mirapaul of Zenith; J. W. Phipps of Thomson Consumer Electronics (RCA); Pat Wilson and Ray Krenzer of Philips Consumer Electronics; and Joe Cagle and Rinaldo Swayne of Alpine/Luxman.

I also wish to thank Joseph A. Labok of Los Angeles Valley College for help and encouragement throughout the years.

And a very special thanks to Daniel Gonneau, Jim Fegen, Larry Jackel, Robert McGraw, Thomas Kowalczyk, Nancy Young, Suzanne Babeuf, Charles Decker, Charles Love, and Jeanne Glasser of the McGraw-Hill organization for having that much confidence in the author. I recognize that all books are a team effort and am thankful that I am working with the First Team.

And to my wife Irene, my research analyst and agent, I wish to extend my thanks. Without her help, this book could not have been written.

LENK'S LASER HANDBOOK

CHAPTER 1
INTRODUCTION TO LASER-BASED DISC PLAYERS

This chapter is devoted to the basics of laser-based disc players, both video and audio. To understand operation of such players, it is essential that you first understand the basics of both TV and stereo systems. If you need refreshers on how TV and/or stereo systems work, read *Lenk's Video Handbook* and *Lenk's Audio Handbook* (Lenk, McGraw-Hill, 1991).

Before we go into any descriptions in this chapter, we take a brief look at the laser-based disc scene, both past and present. With the basics established, we then describe operation of the two most common types of home-entertainment disc players: the *compact disc (CD) audio player* and the *compact disc video (CDV) player*.

1.1 THE LASER DISC SCENE

First, let us resolve the *disc* versus *disk* question. The author generally spells disk with a *k* rather than a *c*. There are those who feel *disc* should be used for consumer audio and video products and *disk* for magnetic data devices. Still others feel that *disc* should be used only with laser recording and playback, and *disk* for other video recording. As a practical matter, most audio and video player manufacturers have settled on disc, so we will do the same.

To further complicate the problems, various manufacturers produce laserdiscs; videodiscs; laservideo, or LV; Laservision; and so on; all of which are the same. That is, the discs have both video and audio information, are played by means of a laser beam, and can be shown on a TV or monitor. In this book, we call the devices on which any form of video discs are played *compact disc video*, or CDV, players. We also discuss *compact disc*, or CD, players, which reproduce only audio (and are generally used with stereo systems or amplifier-speaker combinations).

Note that there are devices that will play both CDVs and CDs. We devote a good part of the book to such players. This is on the assumption that if you understand the operation of and troubleshooting for such a complex player, you should have no difficulty with any other player (CD or CDV). We devote the remainder of the book to the less complex (but far more popular) CD player.

Also note that there are *compact disc interactive* (CD-I) and CD-ROM (read-only memory) devices that use laser-based discs. In fact, the circuits of the CD-

ROM player are quite similar to those of an audio CD player, except for the lack of user operating controls and for one or two additional ICs used to decode digital information. The CD-ROM is used with computers as a substitute for disk drives (both floppy and hard), thus few (if any) operating controls are needed (beyond possible on-off switches, power LEDs, etc.).

We do not dwell on either CD-I or CD-ROM in this book. If you understand the audio CDs, you should have no difficulty with a CD-ROM. A possible exception is the control and interaction between the CD-ROM and the computer. For that reason, we describe interconnections and commands for a typical CD-ROM and computer combination.

1.1.1 The CD Story

CD players are also called *compact audio disc players, digital audiodisc players,* or *simply disc players* in some literature. However, the terms *CD player* and *compact-disc player* are now in the most common use.

We cover CD players designed to reproduce sound from *digital compact discs*, or digital CDs. Such discs are not to be confused with the older analog long playing (LP) records or with pulse code modulation (PCM) records (so-called "digital recordings"). The CDs described here are not interchangeable (electrically or physically) with either LP or digital PCM discs.

1.1.2 The CDV Story

One of the earliest videodisc systems, called TeD, used a pressure pickup (piezoelectric) and was introduced by Teledec Telefunken Decca of Germany in 1975. The TeD system was soon removed from the market. Another system that never quite got off the ground was the *transparent disc* developed by Thomson-CFS.

The *reflective optical pickup* now used for both CD and CDV was developed by NV Philips in the Netherlands and by MCA. The system was called *video long play* at one time and was introduced in the United States in 1978.

A system called CED using *capacitive pickup* was developed by RCA Laboratories and was introduced in the United States in 1981. This was at a time when VCRs were first becoming popular. Since the CED cannot record (as can a VCR), the CED system soon disappeared (although there are some CED players and discs in existence).

Still another system, known as video high density, or VHD, with a companion audio high density, or AHD, was developed by the Victor Company of Japan (JVC) and introduced in 1982. Both systems soon disappeared, primarily because of the VCR popularity.

1.2 INTRODUCTION TO THE CD SYSTEM

The following paragraphs are for readers who are totally unfamiliar with the CD system. A CD player is a very specialized form of phonograph, record player, or turntable. CD players play prerecorded discs (carrying music, speech, etc.)

through a conventional hifi or stereo system (amplifier and loudspeakers). The most common CD is the 5-in disc [the disc is actually 4.75 in (120 mm) in diameter], which can contain between 60 and 70 min of audio. There is also a 3-in CD that contains about 20 min of audio.

Both CDs are single-sided and spin at a high rate of speed compared with conventional audio records. Both use a light-beam and optical-pickup system instead of the stylus and arm found on LP record players. In addition to superior sound (to either analog LP or digital PCM recordings), CD players can provide immediate access to audio at any part of the disc. It is also possible to program CD players to play only selected portions of the audio material.

Figure 1.1 shows the block diagram of a typical CD player. We describe each of the blocks and their functions throughout the remainder of this book. Before we get into such details, let us consider some basic differences between CD and LP players. If you are familiar with conventional record players of any kind, even those capable of reproducing digital recordings, you will see that a CD player is quite different (although the overall purpose is the same).

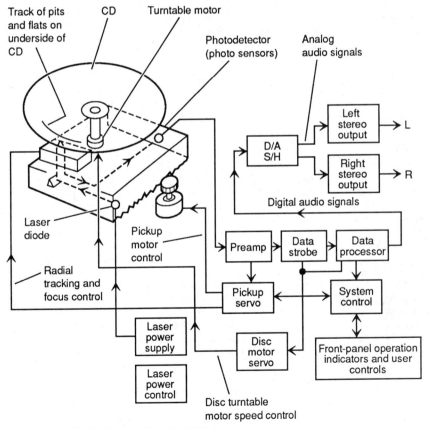

FIGURE 1.1 Block diagram of a typical CD player.

1.2.1 Pickup System and Drive Motors

One basic difference between a phonograph and a CD player is in the pickup. Phonograph records are played with a needle on top of the record. The beginning of the record is at the outside edge, and the needle moves inward as the music is played.

A CD is played from the underside with a light beam. The beginning of the CD is near the center and the light beam moves outward toward the edge as the program plays. The beam is focused up onto the bottom, or underside, of the CD through an *objective lens* (also called the *object lens*), located below the CD (or on the underside of the CD in the case of those players that rotate the CD in a vertical plane).

As the CD is played from beginning to end, the lens is driven by a *servo-operated pickup motor* across the disc. The light beam reflects from microscopic pits on the underside of the CD. These pits are coded with music or other audio, as well as with synchronization and identification data. Note that the lens is part of a *pickup assembly* (sometimes called the *actuator*).

Figure 1.2 shows two basic types of pickups. In one configuration, the optical system (including the objective lens) is mounted at the end of a rotating arm. The arm and lens are rotated (by the servo motor) so that the lens moves from the CD center to the edge. The rotating arm type of pickup has generally been replaced by the *slide-type pickup* (sometimes called the *sled*), which is driven across the CD underside by the motor. While on the subject of motors, there is also a servo-operated *turntable drive motor* in all CD players (to spin the CD) and usually a *loading motor* (to insert and remove the CD from the player).

1.2.2 CD versus LP

Conventional LP record players reproduce the audio signal by tracing the grooves in the record. A CD player reproduces audio by extracting signal information from the disc using a laser beam with no physical contact between the CD itself and the signal pickup mechanism.

The laser beam used to extract the audio is generated by a small, low-power, semiconductor diode, made of aluminum gallium arsenide (AlGaAe), which emits an invisible infrared light. The laser beam is focused onto the CD by the objective lens, which acts like the lens of a microscope and focuses the beam onto a spot slightly less than 1 μm in diameter. The spot is then used to retrieve the information contained on the CD.

Figure 1.3a shows a magnified view of a CD. As shown, the CD is composed of thousands of circular "tracks" made in a continuous spiral from the inside to the outside of the CD. The tracks are similar to grooves in a conventional LP record. However, the tracks on a CD are not true grooves. Instead, CD tracks consist of tiny *pits*, or indentations, in the disc material. The width of the pits is 0.4 to 0.5 μm, and they are 0.1 μm deep. The distance between the spiral tracks is held constant at 1.6 μm, which is called the *track pitch*. (This is not to be confused with *pit spacing*, which is about 2.0 μm from the centerline of one track to the centerline of the next track.) The combination of pits and *flats* (area between the pits) is used to reproduce the recorded information.

Each groove of an LP record contains two signals, one each for the left and right stereo channels, which must be simultaneously read and reproduced by conventional turntable systems. The CD carries left and right channel information

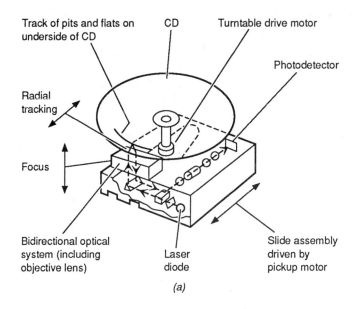

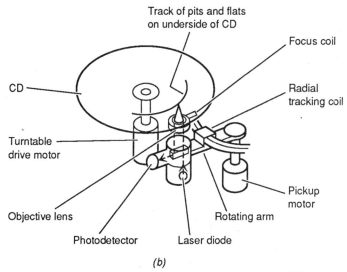

FIGURE 1.2 Two basic types of laser optical pickups. (*a*) Slide type; (*b*) rotating arm type.

separately, with two sets of information aligned successively on the CD. There is a fixed time interval between the two sets of information. As a result, only one information-carrying track is required, and crosstalk between left and right channels is reduced to zero (in theory). In actual practice, channel separation of 90 dB (and better) is quite realistic.

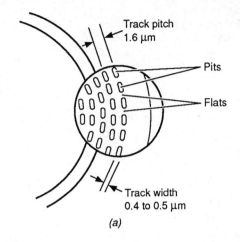

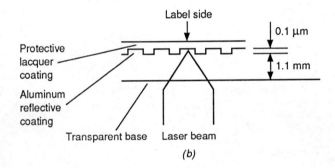

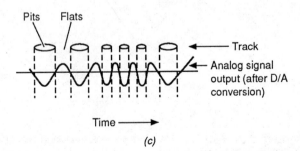

FIGURE 1.3 Magnified views of a CD showing tracks of pits and flats.

In forming signals contained on the CD, the original music or other audio is divided into 44,100 separate signals per second. That is, the original audio is sampled 44,100 times per second, using a sampling frequency of 44.1 kHz. The composition (frequency, level, etc.) of each separate signal sample is then converted into a binary format such as that used in computers (a series of 1s and 0s). This form of storage is called *pulse code modulation*, but it is not to be confused with the PCM of other digital recorders.

The CD signal composition is measured on a scale of 2^{16}, or 65,536, gradations, and the result is expressed as a 16-place number (combination of 1s and 0s). The 16-bit system offers a wider range in which to express the divided signal level than other PCM digital recordings (which generally use a 13-bit system). Theoretically, the dynamic range that can be expressed by the 16-bit system is about 98 dB. However, most CD player manufacturers claim about a 90- to 95-dB dynamic range.

1.2.3 CD Structure and Tracks

As shown in Fig. 1.3*b*, the CD consists of a reflective evaporated aluminum layer covered by a transparent, protective plastic coating. Handling a CD presents far less problems than does handling LP records. For example, even if the CD has some dirt on the transparent base, the laser beam can still operate properly because the beam is directed at the reflective aluminum layer beneath the surface (rather than at the surface). The only possible surface contamination on a CD that can affect playback is something with reflective properties (which can reflect and distort the laser beam). A possible exception is where the dirt completely blocks the beam.

The pits and flats representing the digital information are located 1.1 mm from the transparent surface. The light beam passes through the base material to retrieve the information. The light reflected by the pit is not as bright as the light reflected by the flat area. The CD rotation, combined with the pits and flats passing over the light beam, create a series of on and off flashes of light that are reflected back into the system, thus modulating the light beam.

As shown in Fig. 1.3*c*, the length of the pits and flats determines the information contained on the track. The pits and flats can vary in length from about 1 to 3 μm. The analog waveform shown below the pits and flats represents the decoded signal after digital-to-analog (D/A) conversion. The pits reflect less light than the flat area, and the length of the two vary to recreate the original analog signal.

1.2.4 CD Optical Pickups

Figure 1.4 shows the basic elements of the optical pickup (also called the *optical readout*) used by most CD players with the slide-type pickup. We discuss optical pickups in greater detail in Chap. 2. For now, let us consider the basics.

The laser beam is developed by the laser diode and is applied to the reflective surface of the CD through an optical system (a series of lenses, prisms, gratings, and possibly mirrors, depending on the type of optics). With all three systems, the beam is then reflected back through the optics to a photodiode detector (typically six diodes). The detector produces an output that corresponds to the *audio stored on the CD*. It also produces *tracking* and *focus* signals.

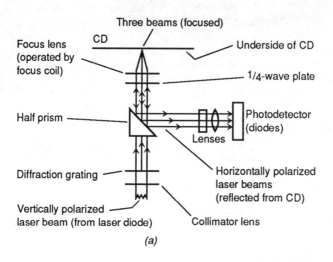

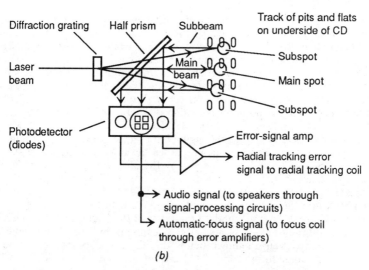

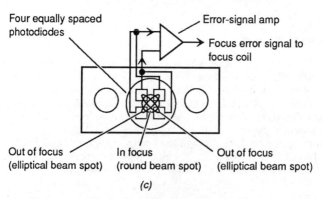

FIGURE 1.4 Basic elements of laser optical pickup.

CD Radial Tracking. With the system shown in Fig. 1.4, the CD spins in a horizontal plane. In many CD players, the CD spins vertically. Either way, it is essential that the laser beam follow the track of pits and flats as the optical pickup is drawn across the CD by the pickup motor, regardless of any possible CD eccentricity. This is called *radial tracking* and usually involves a *radial-tracking coil*, which is operated by a *tracking servo* that moves the lens (and beam) as necessary.

The radial-tracking system uses the three-beam principle shown in Fig. 1.4. This technique uses two subbeams to detect tracking errors and uses the main beam as the audio-signal detector. The subbeams are produced by routing the laser beam through a glass diffraction grating, which creates several images on the same object.

The two subbeams are located ahead of and behind the main laser beam. Also, the subbeams are shifted slightly to the left and right of the main beam. After being reflected by the CD, each laser beam is routed through the optical system to corresponding photodetectors. The error signal from the two subbeams is converted into an electrical signal and then fed to an error-signal amplifier.

As long as CD tracking is precise, the output of the error-signal amplifier is zero. However, if even the slightest radial-tracking error is detected, the input differential between the two error signals (right and left) produces an output. This output is then fed to the radial-tracking servo and coil, which move the objective lens (at right angles to the track) as necessary to correct the position of the main laser beam.

In the rotating arm type of pickup (Fig. 1.2), a coil moves the entire arm and pickup as necessary to restore proper tracking. In most slide-type pickups, the tracking coil moves the optical system (lenses, etc.) in relation to the remainder of the pickup assembly to restore radial tracking.

In a third system, the tracking coil operates a rotary mirror. This mirror is placed between the laser and lens so that the beam makes a 90° turn when reflected by the mirror. The tracking-error signals cause the tracking coil to rotate the mirror (ever so slightly) and direct the beam back to the track.

CD Optical Pickup Replacement. From a troubleshooting standpoint, the exact configuration of the optical pickup lenses, coils, etc., is of little importance. In most present-day CD players, you must replace the entire pickup assembly as a package. You can trace error signals from the pickup diodes, through the servo, and back to the pickup radial-tracking coil. Then you must replace the entire assembly if: (1) the diode signals to the servo are missing or (2) if signals from the servo are available, but the tracking coil does not move the lens (or mirror). About the only replaceable parts on the pickup are the drive motor, drive belt (if any), and possibly some of the drive gears. The laser optics are rarely replaceable (or even adjusted in most cases).

CD Automatic Focus. In addition to radial tracking (keeping the beam centered on the track), the optical pickup also provides for automatic focusing (AF) of the beam to compensate for vertical movements of the CD. This focusing system moves the objective lens (toward or away from the CD) if the laser beam is not focused precisely (within ± 1 μm) on the pits.

The focusing mechanism uses the astigmatism principle. In the simplest of terms, the main laser beam is detected by four equally spaced photodiodes, shown in Fig. 1.4. (These same diodes also reproduce the audio signal.) If the main beam is properly focused, the beam spot is round, and all four diodes receive the same amount of light (and produce signals of the same strength). If the beam is not properly focused, the beam spot is elliptical, and the four diodes receive different amounts of light (and produce different outputs).

The outputs from the four diodes are summed in error amplifiers. The output from the amplifiers represents the focus error. This error (if any) is fed to a focus coil or actuator that moves the objective lens up or down as necessary to correct the focus. The focus coil is similar to a loudspeaker, and the objective lens operates somewhat like a loudspeaker cone.

1.2.5 CD Signal Processing

The block diagram in Fig. 1.1 shows the sequence of signal processing within a CD player. We discuss all of these circuits and more (in fascinating detail) throughout Chap. 5. For now, let us run through the signal sequence quickly.

Preamp and Data Strobe. Since the output signal originating at the optical-pickup photodetectors is very low, the signal is amplified in a preamp stage to a usable level. The signal then enters a data-strobe circuit to discriminate between the 1s and 0s. The data strobe extracts and separates sync signals from the music and other audio signals. These sync signals are encoded on the CD, along with the music, when the CD is manufactured. (The sync signals make it possible for the CD player to reproduce audio at particular points on the CD track, among other functions.)

Data Processor. The next stage is the data processor (or signal processor), which has multiple functions: demodulation of the signal data, error detection and correction, determination of 1 or 0 status, compensation for possible missing parts of the sync signal (performed in conjunction with the data-strobe circuit), random access memory (RAM) control, rearranging data for temporary storage in the RAM, and overall control of the signal-processing circuits.

Interleaving and Deinterleaving. All CD-player circuits include some form of deinterleaving or the rearrangement (unscrambling) of signal data. When a CD is recorded, the music or other audio is interleaved before being recorded on the CD. The interleaving process is especially useful when a relatively large part of the signal is missing. With interleaving, the effects of dropouts in the audio (from any cause) can be minimized.

In the simplest of terms, interleaving involves dividing the audio to be recorded into a series of random sections and then lining up the sections in a new, fixed order before the actual recording. During playback, the sections are rearranged by the opposite process to recreate the original music or other signal. The playback rearrangement processing (deinterleaving) is done by temporarily storing the data in a RAM (usually part of the signal-processing integrated circuit, or IC) and then retrieving the data in the original order.

Thanks to interleaving, even if a relatively large part of the signal is lost, the losses are distributed over various smaller "gaps" in the recreated, final music signal. Because the signals adjacent to the gap are still present, it is easier to compensate for the loss by inserting what are presumed to be the missing parts.

Note that the interleaving or scrambling is done by the recording equipment during CD manufactures and cannot be changed by the CD player. It is the player's function to restore the original signal condition according to information recorded on the CD.

Audio Restoration. A D/A converter follows the signal-processing circuits. The function of the D/A converter is to transform the digital signal back into an analog signal. The converted audio signal is then restored to pure two-channel audio by a sample and hold (S/H) circuit and applied to the left and right stereo output (and to a headphone jack in most CD players).

Servo Control. In addition to the signal circuits shown in Fig. 1.1, most CD players have two servo circuits. One servo controls the CD turntable motor speed (to maintain a *constant linear velocity*, or CLV, as discussed in Sec. 1.2.7) by locking motor speed to signals recorded on the CD. (This function is sometimes called *tangential* tracking and is discussed further in Sec. 1.3.) The other servo controls both radial tracking and the focus of the optical pickup and also controls the pickup motor.

Laser and System Control. In most CD players, the laser has separate laser power-supply and power-control circuits. The system-control circuits shown in Fig. 1.1 control overall operation of the player by accepting commands from the user controls and displaying operating functions on front-panel operation indicators. Generally, the system-control functions are produced by a microprocessor IC, as discussed in Sec. 1.2.9.

1.2.6 CD Audio Reproduction

Figure 1.5 shows the sequence of conversion that occurs within a CD player. The analog waveform A (original music, speech, etc.) is sampled and measured at short intervals as shown in waveform B. The measured values are converted into binary numbers and encoded into a pulse train (also called a *bit stream*) as shown in waveform C. This process involves *quantization*. That is, the maximum signal amplitude that may occur is divided into a number of levels equal to the available number of binary codes. The real value of the analog signal is then rounded to a quantized value that comes closest to the analog value.

The pulse train, or bit stream, shown in waveform C is placed on the CD in the form of pits and flats. The reflected beam is modulated by the pits and flats to create a pulse train of digital information. The detected pulse train is applied to the D/A converter, as shown in waveform D. The detected information is converted back to the original waveform by the D/A converter as shown in waveform E.

To summarize, the audio waveform is sampled at 44.1 kHz, and the value of each sample is measured and converted to a binary number (using quantization). The stream of successive binary numbers is the digital equivalent of the audio waveform. As long as the binary numbers maintain their true values, the waveform is expressed with an accuracy that depends only on the sampling speed of the binary number. The advantage of the binary code, in this respect, is that binary has two conditions, 0 and 1, which can be easily represented by electrical circuits being switched on and off. As long as digital circuits can detect the difference between these two conditions, the stream of numbers is perfectly preserved.

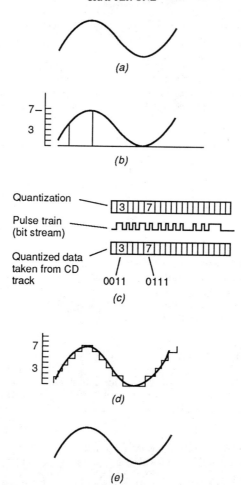

FIGURE 1.5 Sequence of conversion within a CD player. (*a*) Analog input; (*b*) sampling (44.1 kHz); (*c*) A/D conversion (quantization); (*d*) conversion oversampling (176.4 kHz) and filtering; (*e*) analog output.

1.2.7 Constant Linear Velocity

The CD is scanned by the servo-controlled optical pickup at a CLV of 1.3 m/s. To get the scan rate, the rotational speed of the CD is progressively changed from 500 rpm at start-up to 200 rpm at the outside edge of the CD.

The data stream of digital information taken from the CD is kept at a constant rate by a memory. The memory is allowed to fill to half capacity; then data bits are taken from it at the same rate as data comes in, thus maintaining the half-full condition.

If incoming data bits are received at too fast a rate, the memory exceeds the half-full condition and an error signal is developed. This error signal is applied to the turntable motor (through a servo), and CD speed is reduced until the memory remains at the half-full condition. If the CD slows down so that the memory falls below the half-full condition, the error signal polarity is reversed. This causes the turntable motor to speed up, increasing the incoming data-bit rate and restoring the memory to the half-full condition. Disc speed changes are not detected in the reproduced sound since the rotational speed change of the CD has no effect on the rate of speed at which the data bits are removed from memory.

1.2.8 CD Advantages

The most obvious advantage of the CD system is the accuracy of the reproduced sound. If the CD player can tell the difference between a 1 and a 0, the player reproduces the audio exactly as encoded, rounded off to the nearest bit in a range from 0 to 65,536 bits. Not quite so obvious is the ability to insert extra information, or to manipulate the sequence, in a CD data stream. The interleaving and deinterleaving described in Sec. 1.2.5 are examples of this.

These insertions and/or manipulations are done without affecting the original information. This makes it possible to insert automatic error-correcting bits (such as parity bits used in computer systems) into the data-bit stream. Automatic error correction using the added bits can compensate for signal losses resulting from marks or scratches on the CD or from temporary losses in the electronic circuits (dropouts).

1.2.9 CD Player Features

The following are typical features found in present-day CD players (and should be of most interest to technicians facing CD players for the first time).

Front and Top Load. Most present-day CD players are loaded from the front, but many early CD players load from the top. Top-load models are best suited as stand-alone components, but they can also be mounted as top-rack components in an audio system. On most top-load models, the CD compartment cover or lid is opened by a push button but must be closed by hand. Top-load CD players do not require a *loading motor* and are thus simpler than front-load models.

Front-load CD players can be used as stand-alone components or can be operated at any location in an audio-system rack. There are two basic versions of front-load models: horizontal and vertical.

With horizontal front loading, you press a front-panel button to open a drawer or tray, insert the CD, and close the drawer (manually) or tray (with a control). In the drawer version of horizontal front loading, the entire turntable and optical pickup are mounted in a drawer and are moved in and out of the front panel. With this system, you operate a front-panel control, the CD drawer slides out automatically, and a CD pressure plate raises. You then position the CD on the turntable and push the drawer back into the player. This pulls the pressure plate over the CD and places the player in a ready-to-play condition.

Note that, in general, the top-load and drawer-type front-load players use the rotating arm pickup (Fig. 1.2*b*), whereas the horizontal-tray and vertical front-load players use the slide-type pickup (Fig. 1.2*a*). The present trend is for horizontal-tray design in all CD players, even those with multiple CD capability.

With the tray version of horizontal front-load, the turntable and pickup are in the player, and the CD is inserted and removed by means of the front-panel tray. The tray is operated by a loading motor in response to a load-unload (or open-close) control. One touch of the control moves the tray out to a position where the CD can be inserted (or removed). Another touch of the control causes the tray to be pulled in (and to position the CD over the turntable).

With vertical front loading, the turntable and pickup are in the player. A loading motor opens and closes a vertical door (hinged at the bottom) so that the CD can be inserted and removed.

On virtually all CD players, there are circuit breakers and safety switches (interlocks) that prevent operation of drive motors and the laser when the CD drawer or tray covers are open. We discuss these interlocks and switches, and the mechanism they control, in Chaps. 5 and 7, respectively.

Random Memory Programming. With a CD player capable of random memory, you can preset up to 15 or 20 programs (individual selections on the CD) for playback in any order. Typically, you enter the number of the desired programs by first pressing a Program button; then you use the buttons marked 1 and 10. If the first program you want is the third program on the CD, you press the Program button and enter the number 3 by pushing the 1 button three times. The program indicator displays the number 3 for confirmation. Pushing the Play button after the full sequence has been entered starts playback of the program sequence entered.

Self-Program Search. During playback, CD players with a self-program option let you skip forward and backward to locate the beginning of each program on the CD. In a typical CD player, you press the FF (fast forward) control once, and the optical pickup advances to the beginning of the next program (and begins playing the CD at that point). When you press the FB (fast backward) control once, the pickup moves back to the beginning of the current program to begin play. If you press the FB control twice, the pickup moves back to the beginning of the previous program on the CD to begin play.

CD Scanning. On players equipped with scanning, the player is placed in the scan mode when the play control and the FF or FB control are operated simultaneously. This causes a brief sample of the current program to be played. Then the pickup advances to a point about 30 s ahead (or behind) the CD play time, and another brief sample is played. This process continues as long as the FF or FB controls are engaged.

Memory Stop. On CD players with memory stop, you can mark any point on the CD for instant location with the FB control. In either play or pause modes, you mark the current CD location (the beginning of a favorite program, for example) by pushing the memory stop control. The point can then be returned to while in play or pause modes by pushing the FB control. The pickup moves back to the memory-stop location, and the player automatically goes into the pause mode. You then push the Play control to start play from the memory-stop point.

INTRODUCTION TO LASER-BASED PLAYERS 1.15

Repeat Play. Play of the entire CD, or play of a random memory programming sequence, can be repeated continuously on most CD players.

For play of the entire CD, you push the repeat control at any point prior to or during play of the CD. After the full CD is played, the pickup returns to the beginning and begins play again. Repeat play of a random memory programming sequence is generated in the same way.

Display Callouts. On some CD players, there is a call control which determines what display is shown on the front-panel indicators. Typically, there are indicators on the front panel that show such conditions as total CD playing time, elapsed playing time, number and total time of programs entered (via random memory programming), and possibly the track or index numbers being played. We go into typical operating procedures for CD players in Chap. 3.

Microprocessors. As in the case of most present-day electronic equipment, CD players are under the control of at least one microprocessor. This makes possible all of the features described thus far. We do not go into all features here, since such functions are described in Chaps. 5 and 7. However, it is important to note that one major function of the microprocessor is to provide for random memory programming in conjunction with the disc directory.

Disc Directory. CDs are digitally encoded at the beginning of the program material so the player will know the length of the program. Also encoded at the beginning of each selection is an individual code that identifies the location of that particular selection. This system of identification (called the *disc* or *CD directory*) allows each selection on the CD to be accessed using the microprocessor and random memory programming feature.

Typical CD Player Specifications. The following specifications are for a "typical" CD player and are included here for reference:

    Audio
        Number of channels: 2
        Frequency response: 5 to 20,000 Hz ±0.5 dB
        Dynamic range: 93 dB
        Signal-to-noise ratio: 94 dB
        Harmonic distortion: 0.003 percent (at 1 kHz)
        Channel separation: 92 dB (at 1 kHz)
        Wow and flutter: Less than measurable limits (crystal controlled)
        Output voltage: 2.1 V rms (with full scale; 0 dB into 50 k)
        Headphone output: 20 mW (0 dB in 8 Ω)
    Signal format
        Sampling frequency: 44.1 kHz
        Quantization number: 16-bit linear/channel
        Transmission bit rate: 4.32 Mb/s
    Pickup
        System: Objective lens drive system (optical pickup)
        Lens drive system: Two-dimensional parallel drive
        Optical source: Semiconductor laser
        Wavelength: 7900 Å

Discs used
  Playing time: 60 to 70 min on one side
  Diameter: 120 mm
Functions
  Random memory search (15 tracks), self-program search system, skip play, memory stop, pause, repeat, output volume adjust
Displays
  Play position (5-min steps), number of tracks, playing track number, elapsed play time, output volume level
Outputs
  Two sets of output terminals (variable and fixed level) on rear panel
  Variable-level headphone jack on front panel
Accessories
  Connecting pin cords for connection to stereo system
  Demonstration CD (part of the user operating instruction package)

1.3 INTRODUCTION TO THE CDV SYSTEM

The following paragraphs are for readers totally unfamiliar with the CDV system. As in the case of CD, the CDV system uses discs on which information is recorded in the form of pits and flats. The pit and flat track format shown in Fig. 1.3a is the same for both CDs and CDVs. However, much more information is required for CDVs since both audio and video are involved.

Figure 1.6 shows the block diagram of a combination player for both CD and CDV discs, while Fig. 1.7 shows the frequency spectrum for CDV. As shown, the output of the player is both audio and video.

When the player is used for standard 8- or 12-in videodiscs (Laservision discs,

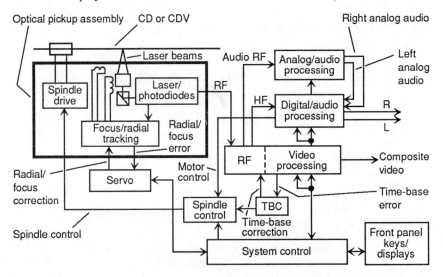

FIGURE 1.6 Block diagram of a combination player for both CD and CDV discs.

INTRODUCTION TO LASER-BASED PLAYERS

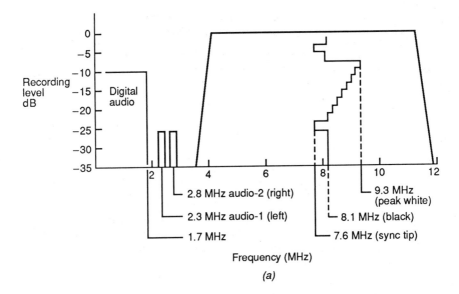

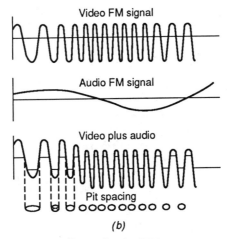

FIGURE 1.7 (*a*) Frequency spectrum; (*b*) encoding for CDV.

or LD), the composite video and left-right analog audio are output to a color monitor or TV. When the player is used with 3- or 5-in CDs, only the digital audio is output to a hifi stereo system. The player in Fig. 1.6 can also be used with the newer 5-in discs (so-called *Gold CD* or *CDV Single* discs) that contain both digital and analog audio, as well as composite video.

1.3.1 CDV Frequency Spectrum and Encoding

As shown in Fig. 1.7a, the composite video signal is frequency-modulated with a deviation of 1.7 MHz, from 7.6 MHz (TV sync tip) to 9.3 MHz (at TV white peak). The TV black level is located at 8.1 MHz.

The left analog audio channel (audio 1) is frequency-modulated on a carrier at 2.3 MHz, while the right channel (audio 2) uses a carrier at 2.8 MHz. Both analog audio carriers have a maximum deviation of 100 kHz. The digital audio is pulse-width modulated (PWM) at frequencies below 1.7 MHz.

The signals are summed as shown in Fig. 1.7b. (Although four signals are involved, only two signals are shown.) The combined signal results in the PWM signal used to etch the pits onto the disc. The disc encoding process is discussed further in Chap. 2.

1.3.2 CDV Control Codes

Control codes are included with the composite video and audio information. These codes are placed on TV horizontal lines during the vertical blanking interval, as shown in Fig. 1.8. For example, line 17 of the first TV field contains a digital code which represents the picture (frame) number for a constant angular velocity (CAV) disc or the elapsed time for a CLV disc (Sec. 1.3.3). Line 280 contains a code which represents the chapter. Lines 18 and 281 are duplicates of lines 17 and 280, respectively, which provide backup if the previous lines are lost because of dropouts. Other data may be included on other lines during the vertical blanking interval, such as stop codes for interactive discs.

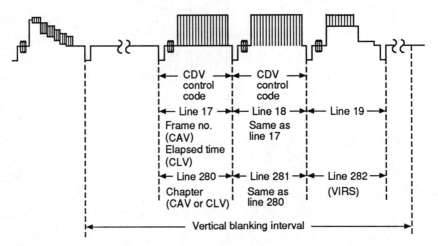

FIGURE 1.8 CDV control codes placed on TV horizontal line during the vertical blanking interval.

1.3.3 Disc Formats (CLV and CAV)

In addition to the variation of sizes, there are two different videodisc formats for placing the video information on the disc. The two formats are the *extended play*, or CLV, disc, and the *standard play*, or CAV, disc. Figure 1.9 shows the relationship among the disc sizes, while Fig. 1.10 shows the CLV and CAV formats for the 12-in disc.

Standard Play CAV. Each revolution of a standard play CAV disc contains one frame of video information (Fig. 1.10a). One TV frame consists of two interlacing fields. Thus, the TV screen is scanned twice for each revolution of the CAV disc.

Each field is separated by the vertical blanking interval, which becomes longer as the interval extends to the outer circumference of the disc. This format allows the disc to be played at the same speed of 1800 rpm, which is equivalent to the National Television System Committee (NTSC) frame rate of 30 Hz (1800 ÷ 60 = 30) throughout the disc.

The CAV format allows the disc to be interactive and provides for special playback effects such as still, fast slow, and strobe. The disadvantage of CAV is that the video content is limited to 30 min per side on a 12-in disc (54,000 frames per side divided by 1800 rpm equals 30 min). The 8-in CAV disc contains about 14 min of video per side.

Extended Play CLV. Unlike CAV, CLV discs do not maintain the constant arrangement of vertical fields. Instead, the field is of equal length throughout the diameter of the disc, as shown in Fig. 1.10b. CLV allows the signal information to be packed more densely on the disc, which extends the play time of a 12-in disc to 1 hr per side (108,000 frames). The 8-in CLV disc contains about 20 min of video per side. In general, CLV also results in reduced signal-to-noise (S/N) ratio and frequency response (because of the denser packing of signal information.

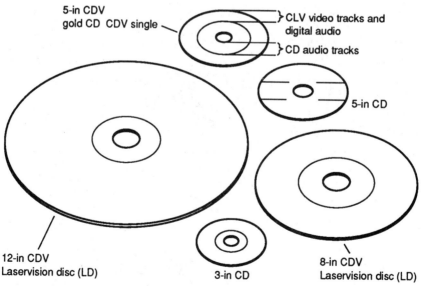

FIGURE 1.9 Relationship among disc sizes.

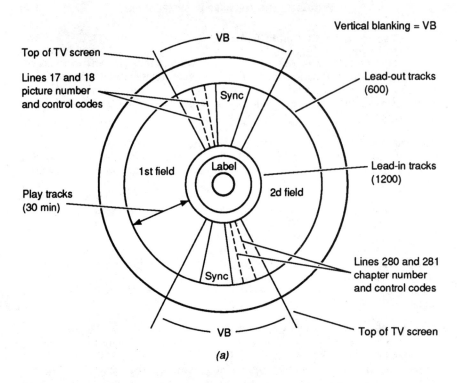

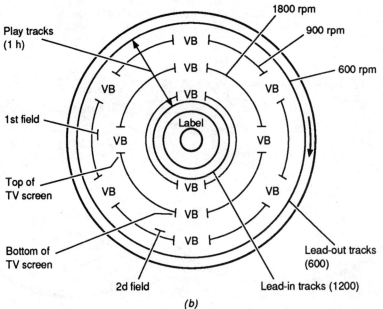

FIGURE 1.10 CLV and CAV formats for the 12-in disc. (*a*) Standard play or CAV disc (motor speed constant; vertical field track length variable); (*b*) extended play or CLV disc (motor speed variable; vertical field track length constant).

As a CLV videodisc plays, the rotation speed of the disc is reduced when the pickup laser beam tracks from the inside to the outside circumference (as discussed for the CD system in Sec. 1.2.7). The velocity of the track passing the pickup beam is the same over the entire disc. Typically, videodisc speed starts at about 1800 rpm when tracking closest to the center of the disc. Toward the outer diameter, disc speed changes to about 500 rpm.

1.3.4 Lead-In and Lead-Out Tracks

Both CAV and CLV videodiscs have lead-in and lead-out tracks. Lead-in tracks are located prior to the start of program material and contain a *start code* used by the player to move the pickup beam to the start of material. Lead-out tracks (typically a minimum of 600 tracks on both CAV and CLV) are located at the end of program material and contain an *end code* to identify the end of material. On most videodiscs, there is also a *lead-out code* that instructs the pickup beam to return to the start of the disc.

1.3.5 Gold CD and CDV Single

As shown in Fig. 1.9, the Gold CD or CDV Single disc (5-in) contains video tracks in the CLV format, as well as standard CD audio tracks. The video tracks, with accompanying audio, are located toward the outer circumference of the disc. The CD audio tracks are located at the inner circumference. The diameter of the Gold CD is much smaller than the conventional 8- or 12-in videodisc. The video information is thus contained within a smaller radius.

The outer tracks of the Gold CD are located about where the inner tracks begin on the 8- or 12-in disc. As a result, the Gold CD must spin faster than the larger discs. The speed of the video portion of the Gold CD is from 2700 to 1800 rpm. The CD audio portion speed is from 500 to 300 rpm. Note that the audio portion of the Gold CD may also be played on any standard CD player (Sec. 1.2) since it is identical to the standard CD.

1.3.6 Basic CDV Operation

By comparing Fig. 1.6 with Fig. 1.1, you will see that operation of the CDV player is similar to that of the CD player described in Sec. 1.2. The following paragraphs summarize both the differences and similarities.

Laser Pickup. Like CD, the CDV player reads prerecorded information with a noncontact optical pickup system using a low-power laser beam. The beam spot (with a diameter of 1 μm) strikes microscopic pits as the disc spins. This creates reflected light that varies in intensity according to the spacing and length of the pits. The laser beam is locked on the track of the pits by a servo-controlled system to keep the laser beam on track and in focus. Disc rotational speed is also controlled by the servo.

Advantages and Features. As in the case of CD, the optical pickup used by CDV players provides important advantages and playback features. We discuss typical features in Sec. 1.3.7.

Chapters and Frames. Most CAV videodiscs contain segments, or chapters, as well as numbered frames (pictures) which may be displayed on the monitor or TV. Usually, the number of the information being displayed appears on the TV screen (on-screen display, or OSD) and on the front panel of the player.

RF and Control-Code Processing. RF from the laser pickup is sent to the video processing circuit where information read from the disc is converted to composite video. Control codes recorded on the disc are decoded and used by system control to operate the player and to provide display information to the front panel and OSD (for display on the monitor or TV screen).

Time-Base Correction (TBC). Video is also fed to the TBC circuit to provide timing information to the spindle-control circuit. In turn, the spindle-control circuit provides speed correction and start-up signals to the spindle (disc turntable) motor through the spindle-drive circuit. The TBC also provides time-base correction to the video signal before video is sent to the monitor or TV.

Video Processing. The video-processing circuits send audio RF to the analog audio-processing circuits for demodulation of the FM analog signal. The video-processing circuits also send high-frequency (HF) signals to the digital audio-processing circuits for decoding.

1.3.7 CDV Player Features

The following are typical features found in present-day CDV players (and should be of most interest to technicians facing CDV players for the first time).

Disc Compatibility. Early CDV players are generally limited to 8- and 12-in videodiscs. Present-day CDV players can also play 3- and 5-in CDs, as well as the newer Gold CD and CDV Single discs.

Displays. Virtually all present-day CDV players provide for on-screen display of chapter and frame information on the monitor or TV. This same information is also displayed on the player front panel (usually with a fluorescent tube, but with an LED display in some cases).

Special Effects. Most CDV players can provide both multispeed and still and step play.

Audio Features. Virtually all present-day CDV players have dual 16-bit digital-to-audio converters (DACs) and provide for *CX noise reduction*. The CX noise-reduction system serves to cut down noise by about 10 dB or more without compromising the frequency response. CX thus expands the dynamic audio range and enables reproduction of audio with a good S/N ratio.

Note that when discs recorded with CX-encoded audio are played, it is necessary to press a CX button to activate the CX noise-reduction circuits. This sets the encoded audio signal to a normal level, but the noise is reduced. Also note that playing a normal (nonencoded) disc with the CX system on, or playing a CX-encoded disc without the CX system on, *results in unnatural reproduction of the audio.*

Most CDV players provide 4× oversampling of the audio and have stereo headphone jacks (with separate volume control). Many players provide for random play as well as favorite-track selection (FTS) of the CD audio.

Miscellaneous Features. Most CDV players can be operated by remote [using an infrared, remote (IR) system] and are capable of repeat play as well as chapter and track programming (with 15 or 20 selections). The most sophisticated (and expensive) players provide for S-VHS or S-Video outputs. Typically, a CDV player provides 425 lines of horizontal resolution.

Digital Special Effects. The most sophisticated CDV players provide a number of special effects (using the remote control) for CAV, CLV, and CDV Single discs (but not for CD audio discs). Such effects are described further in Chap. 3. Typical effects include still and step, multispeed, search, skip, picture freeze, strobe, memory, and jog (for continuous adjustment of playback speed).

CHAPTER 2
ENCODING, DECODING, AND OPTICAL READOUT

In this chapter we discuss the process involved in encoding the original audio and video onto the CD or CDV in the form of pits and flats and then decoding the tracks of pits and flats back to audio and video suitable for reproduction on a monitor, TV, and/or stereo system. Remember that the encoding takes place at the time of disc manufacture and cannot be altered. It is the job of the player to decode the disc signals. However, to understand operation of the player decoder circuits, it is most helpful (if not essential) to understand the encoding process. We also describe additional optical readout principles not covered in Chap. 1.

2.1 ENCODING A CD

Figure 2.1 shows a simplified block diagram of the encoding process, which is typical for the equipment used in CD recording. To produce a CD, the audio signals are digitized and encoded before any other processing takes place. The first step in the process involves the use of low-pass filters (LPFs).

2.1.1 LPFs

Both left and right input audio signals are passed through sharp-cutoff LPFs, as shown in Fig. 2.1a. These filters limit the bandwidth to a maximum frequency (f_m) equal to (or less than) the sampling frequency (f_s) of 44.1 kHz. If the audio frequency to be sampled is greater than the sampling frequency, intermodulation distortion can occur because of *frequency foldover*, as shown in Fig. 2.1b.

2.1.2 Sample and Hold (S/H)

Before the stereo signals are recorded on the CD, the signals are converted into a digital format. One of the first conversion steps is to sample the audio signals at fixed intervals, or *time points* (also called *sampling points*). This sampling is done by the S/H circuits shown in Fig. 2.1a. Figure 2.2 shows the waveforms produced when a single cycle of audio is sampled.

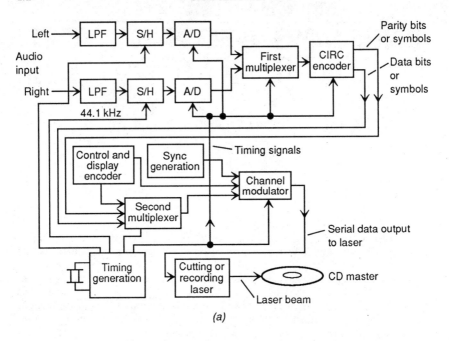

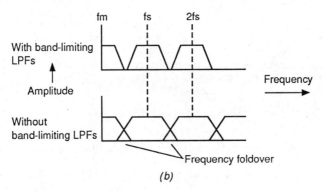

FIGURE 2.1 CD encoding process.

In addition to sampling, the S/H circuits also measure the values for each sample of the audio signal, as shown in Fig. 2.2b. The measured value is held for a moment to permit conversion to a *binary-coded waveform*, as shown in Fig. 2.2c. The maximum audio signal frequency that can be sampled in this way is *one-half the sampling frequency*. The sampling rate of 44.1 kHz is more than sufficient for the typical audio range of 0 to 20 kHz.

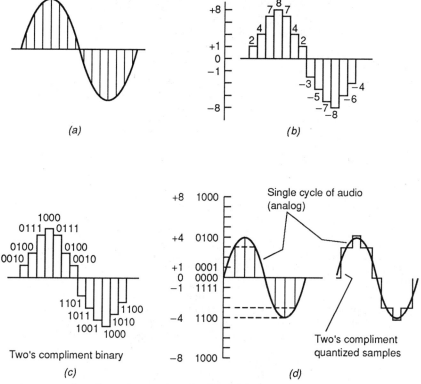

FIGURE 2.2 Sampling and conversion of audio waveforms.

2.1.3 Analog-to-Digital (A/D) Conversion

After sampling the analog signal, the next step is to convert each sample into a two's complement binary code, as shown in Fig. 2.2d. One of the problems with this process is that audio signals can assume an infinite number of levels, whereas the number of binary codes available to reproduce the levels are finite. To overcome this problem, the sample signal is *quantized*. That is, the maximum value that occurs is divided into a number of levels equal to the available number of binary codes. This process occurs in the A/D converters, which produce a number of binary bits representing the quantized level of each sample.

2.1.4 Multiplexing

Sampling, quantization, and conversion to binary code are operations performed separately for the left and right stereo channels. However, both channels are ultimately stored on the CD, *one after the other on a single track*. To accomplish this, the first multiplexer circuit is used to pass the information, in sequence, to the Cross Interleaved Reed-Solomon Code (CIRC) error-correction encoder circuit.

2.1.5 Error Correction (CIRC)

The error-correction circuit processes the digital signal (binary bits from the multiplexer) by a method called CIRC, which involves both parity bits and interleaving. As shown in Fig. 2.1a, there are two outputs from the CIRC encoder. One output carries data and the other carries parity bits. The two outputs are applied to a second multiplexer which, in turn, feeds the signals in sequence to the channel modulator. It is at this point in the process that the control and display encoding is mixed with the data and parity bits.

2.1.6 Control and Display Encoding

The control and display encoding provides operating features for the CD player which cannot be found in conventional record players. At the beginning of each CD is encoded the number of selections on that CD. Each musical selection is identified separately so that the selection can be readily accessed.

Identification codes can also be used for other purposes. For example, the codes that identify the pause between two musical selections can be used to identify whether or not the recording is made with preemphasis. (This makes it possible to switch the deemphasis circuits of the player on or off automatically.) Timing information can be encoded for displaying elapsed time or the length of time a particular piece of music has played.

Control and display information is nonaudible and is encoded separately. The control and display encoder outputs 8-bit symbols, permitting the implementation of eight different information channels. The control and display information is usually referred to as *subcoding* and is added to the data and parity bits in the second multiplexer.

2.1.7 Sync Generation

The output of the second multiplexer is fed (serially) to the channel modulator, with the data arranged in blocks. In order to recognized the blocks of data, the *sync generator* is used to generate a unique pattern (which is not contained in the normal data). The sync pattern is passed to the channel modulator, which adds the pattern to the data at intervals based on timing pulses received from the timing generator.

2.2 CHANNEL MODULATION (CUTTING THE MASTER)

Now that we have gone through the basic encoding process, let us go back through the encoding circuits using some typical audio signals. We will see how the circuits in Fig. 2.1 are used to modulate the cutting or recording laser to produce a corresponding track of pits and flats on the CD master (used to manufacture CDs).

The information created from the two stereo channels during the sampling period is used to convert the signals to a binary code. The code is then used to control the cutting laser by modulating the laser beam during the manufacture of the CD master. Once the master is cut, the CDs are produced by a stamping process.

Note that the right-channel signal is somewhat lower in amplitude than that of the left channel and is almost twice as long (lower in frequency). However, both L and R channels are sampled six times. This configuration is called a *frame*, as discussed next.

2.2.1 Frame Organization

As discussed in Chap. 1, a CD is composed of billions of tiny pits, which represent the digital encoded data. In order for the CD player to recognize this data, it is necessary to organize the data into patterns. In all CDs, the pattern is organized into a block structure, which is the frame. The frame is a period of time that contains six audio samples (of both channels).

The six audio samples of the frame are made up of six sample periods of both left- and right-channel audio so as to equal 12 sample periods. The two channels of audio are applied through separate LPFs to the S/H circuits. In turn, the S/H circuits measure (or sample) the voltage level of the analog signals in short (22.7 μs) intervals at a frequency of 44.1 kHz. This measurement is then converted into a binary code by the A/D converter.

The output of the A/D converter is a 16-bit binary code representing that sampling period of the audio signal in digital form. Note that the 16-bit code is in two's complement form (Fig. 2.2d) to accommodate both positive and negative swings of the audio signal. Also, in most CD literature, each 16-bit number is called a *word*, and each word is split into two *symbols* of 8 bits each.

No matter what it is called, conversion of the audio signal to a 16-bit binary code is done separately (from each other) by two channels containing the same circuitry (and synchronized to each other by the timing generator). The two channels of data (each containing six sample periods of sequential data) are applied to the first multiplexer, which switches the two data streams into one sequential data stream.

The output of the first multiplexer is one single data stream containing the 12 sample periods (words) of both L and R channels intermixed with each other so that the L sample period is followed by an R sample word. The sequential data stream from the first multiplexer is applied to the error-correction circuit, where interleaving takes place.

2.2.2 CIRC Interleaving

One function of the error-correction circuit is to break each 16-bit word into two 8-bit symbols. During this process, the two 8-bit symbols are interleaved (rearranged or scrambled). After interleaving, the next step is to provide a method by which the CD player can check for errors in the data stream. This is done by parity bits generated in the error-correction circuit. The parity bits are generated as 8-bit data words and are called a *parity symbol*.

2.2.3 Parity Symbol

Two channels of information are produced by the CIRC error-correction circuit, as shown in Fig. 2.1a. One channel contains the data symbols representing the audio samples; the second channel represents the parity symbols. The data and parity symbols are blocked out of the error correction to the second multiplexer at intervals determined by the timing generator. The multiplexer adds the parity symbols to the correct location in the data stream so that the decoder in the CD player can check for errors in the data symbols.

2.2.4 Subcode Data Symbol

The subcode data symbol (often called the subcode) not only provides identification of the entire frame of data symbols but also provides a *sync pattern* used by the CD player to process the data symbols. The subcode is processed by a microprocessor in the player, and it provides the system with updated information as to where the optical pickup is located (in the pickup's travel across the CD). This information is displayed on the front panel of the player. The subcode also provides control information to aid the microprocessor in locating that particular frame on the CD. The output of the second multiplexer, which contains the subcode (8 bits) 24 data symbols of audio (also 8 bits each) and 8 parity symbols (again 8 bits each), is applied to the input of the channel modulator.

2.2.5 Channel Modulator and EFM

The frame is applied to the channel modulator which, in turn, provides modulation to the cutting, or recording, laser. Note that the data stream is not placed directly on the CD in the 8-bit format. Instead, the 8-bit code is converted into a 14-bit code called *eight-to-fourteen modulation*, or EFM. As the name implies, each group of 8 data bits is converted into a 14-bit data symbol. In effect, the process "stretches" the distance between adjacent pits and flats, as shown in Fig. 2.3. The stretching from 8 to 14 bits prevents the laser beam (in the player) from converting two adjacent transitions (from a pit to flat and vice versa) at the same time.

There is no direct relationship between an 8-bit and a 14-bit code in their natural state. So, to make the two codes compatible with each other, two conditions are placed on the 14-bit code. First, there must always be at least two 0s between successive 1s. Second, there can be no more than 10 consecutive 0s in a 14-bit run of the code. When these two conditions are applied, all but 277 possible combinations are eliminated. The 277 codes remaining can be further reduced by eliminating 21 codes that contain the longest run of 0s, which leaves 256 possible combinations. An 8-bit code contains 256 combinations ($2^8 = 256$), so the two codes (8-bit and conditioned 14-bit) have a direct one-to-one relationship.

The EFM process is reversed in the CD player to provide correct decoding. Typically, this is done by placing the 8-bit code in a lookup table in a ROM (which is part of the CD player's decoder integrated circuit (IC), as discussed in Chap. 5).

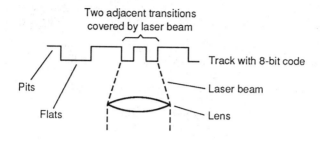

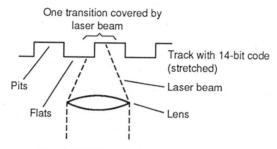

FIGURE 2.3 Effect of EFM conversion.

2.2.6 Symbol Sync and Frame Sync

In addition to EFM conversion, the channel modulator adds a unique sync pattern, created by the sync generator, to the EFM data stream. The sync pattern generated by the subcode (Sec. 2.2.4) is used to process the data symbols during error correction and is called *symbol sync*. The 24-bit sync pattern that is placed at the beginning of the frame is used to provide clock and timing signals for the player decoding system and is called *frame sync*.

Merging bits are added to the end of each symbol. Without the merging bits, the symbols would violate the two conditions placed on the 14-bit code since the symbols are linked together serially. The merging bits contain no audio or other information and are therefore skipped by the decoding system in the CD player.

2.2.7 Conversion to Pits and Flats

The output of the channel modulator provides the data stream to the control circuit of the cutting laser. The signal modulates the laser beam, thus cutting bits into the master (sequentially, in a spiral pattern from the beginning to the end of the master).

2.2.8 Subcode Information

At the beginning of the encoding of a master, the subcode contains information indicating the *number of selections* recorded on the CD. This information is encoded in the same frame structure as all other data, except that the subcode data symbols in the frame contain no audio information. This is done by placing all zeros at the subcode data symbols.

The subcode also contains a bit (known as the *pause bit*) to provide information as to when a selection is about to begin. The pause bit is placed at the beginning of each selection of music in an area containing no audio and is used to tell a microprocessor in the CD player that a selection is ready to begin. This allows the microprocessor to read the subcode and determine if the selection is correct (as determined by the front-panel controls).

2.2.9 System Frequency

The frequency of the CD system is based on the six audio samples that contain the 12 sample periods of left- and right-channel audio. There are 588 channel bits in one frame. The frame can then be divided into six audio samples, which equals 98 channel bits per audio sample. The 98 channel bits are then multiplied by the sampling frequency of 44.1 kHz to produce a system frequency of 4.3218 MHz. Note that the system frequency is sometimes referred to as *high frequency*, or HF, in CD player literature.

2.3 DECODING THE CD

Figure 2.4 shows the simplified block diagram of the decoding process. If you compare this to the block diagrams in Chaps. 1 and 5, you will see that Fig. 2.4 is essentially a block diagram of the complete CD player, but the focus and radial-tracking functions have been omitted.

As shown in Fig. 2.4, the HF signal retrieved from the CD by the optical pickup is amplified and filtered in the high-frequency amplifier. The output of the amplifier is a frame-structured data stream containing the EFM format. The amplified HF signal is applied to the EFM demodulator, which forms the front end of the decoding system.

The EFM demodulator supplies the demodulated data and timing signals to the error-correction (ERCO) circuits, as well as to the control and display decoder. The HF signal is also applied to the clock regeneration and synchronization detecting and timing circuits to recover the bit clock and sync pattern from the data stream (which, in turn, provides timing for the system).

2.3.1 Error Correction during Decoding

The error-correction IC performs the task of error detection and correction. This information is supplied to the interpolation and muting IC, together with a flag signal. The ERCO compares the derived clock signal to a reference frequency (crystal controlled at 4.3218 MHz) to detect any discrepancy in the data stream. If an error is detected in the data stream by a comparator in the ERCO, an error-correction signal (the motor-control error signal) is applied to the turntable motor servo amplifier.

If the data rate coming into the ERCO is less than 4.3218 MHz, the error sig-

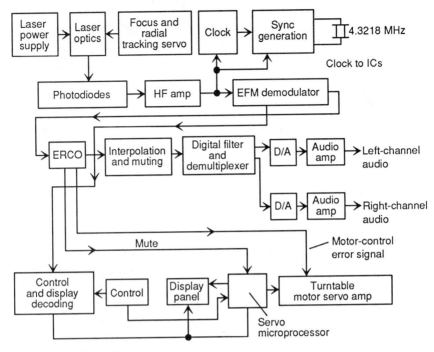

FIGURE 2.4 CD decoding process.

nal increases the turntable motor speed, and vice versa. The motor-control error signal controls CD speed as play continues from the inside of the CD (about 500 rpm) to the outside edge (about 200 rpm), thus maintaining constant linear velocity (Sec. 1.2.7).

2.3.2 Interleaving and Deinterleaving

The majority of errors that may occur during playback of a CD results from scratches, dust, and dirt that may reflect the laser beam (producing erroneous bits). Because of the high density of information on CDs, such defects or reflections can easily wipe out several adjacent bits or samples on the track. If all the affected samples belong to the same frame, a great many *multiposition errors* can occur inside each frame. Interleaving is used to avoid multiposition errors in a frame during playback.

Interleaving is based on the fact that analog signals are continuous signals that usually do not change abruptly. The amplitude of the signal during the first sample does not differ greatly from that of the second sample. The amplitude during the third sample does not differ much from the second sample, and so on. If the value of the second sample is lost, but the value of the first and third samples is known to be good, then an approximation or interpolation can be made to compute the second sample.

The basic principles of interleaving and deinterleaving are shown in Fig. 2.5. Remember that interleaving is done by means of delay lines (having different de-

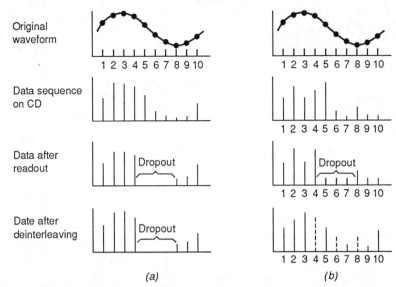

FIGURE 2.5 Basic principles of interleaving and deinterleaving. (*a*) Without interleaving; (*b*) with interleaving.

lay times) allocated to specific samples during the encoding process (at the time of CD manufacture). Deinterleaving occurs in the CD player at the time of playback.

In Fig. 2.5*a*, the sequence of signal processing is shown without interleaving. The audio signal is first sampled at time points 1, 2, 3, and so on, and is digitized and recorded onto the CD. If there is a dropout during playback of the CD, there are some symbols missing in the received data. In the example used, three symbols (5, 6, and 7) are missing, producing a serious dropout.

In Fig. 2.5*b*, the same sequence of signal processing is shown with interleaving. Again, the audio signal is sampled, but with the samples rearranged prior to CD recording. Such interleaving results in the recording of data in a sequence which does not represent an increasing time scale. During reading of the CD, the same dropout occurs and again results in three missing data symbols (4, 6, and 8). Deinterleaving is then performed in the CD player to restore the original sequence of data symbols.

It can be seen that with interleaving there are no multiposition errors and that the single missing data bits can be approximated by interpolation. For example, symbol 4 can be interpolated as a value between symbols 3 and 5 (less than symbol 3 but greater than symbol 5).

2.3.3 Digital Filter and Demultiplexer

The deinterleaved and interpolated data stream is applied to the digital filter and demultiplexer IC where the stream is filtered and separated into left- and right-channel data. Digital filtering provides a higher reproduction quality than is found in conventional record players. However, to get this quality, *the sampling signal frequencies must also be filtered out.* If the series of binary numbers representing quantized samples are simply converted to analog values, a series of *sampling spikes* can occur, as shown in Fig. 2.6a.

Although the outline of the spikes resembles the audio waveform, the constant on-off switching produces an infinite series of frequencies. To overcome this problem, each sample is held until the next sample arrives. This produces a step waveform, as shown in Fig. 2.6b. The step waveform is much nearer the shape of the original audio waveform than is the series of spikes. The S/H function occurs in the digital-to-analog (D/A) converters.

2.3.4 D/A Converters and Oversampling

D/A conversion is the last step in the processing sequence before stereo audio amplification. The right and left D/A converters transform the 16-bit digital code into an analog signal that has the same shape as the original audio signal. Simultaneously, the converters use oversampling, which acts as a preliminary filter, to refine the step waveform shown in Fig. 2.6c to a step waveform resembling that in Fig. 2.6d.

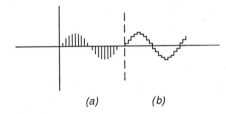

(a)   (b)

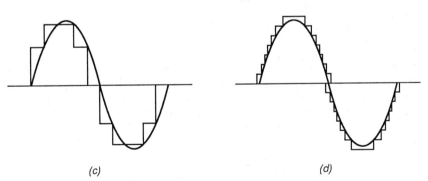

(c)   (d)

FIGURE 2.6 Effects of sample-and-hold, oversampling, and digital filtering. (a) Sampling spikes; (b) step waveform; (c) audio waveform sampled at 44.1 kHz; (d) audio waveform sampled at 176.4 kHz.

The oversampling, or preliminary filtering, is done by digital filters that operate at 4 times the sampling frequency (4 × 44.1 kHz = 176.4 kHz). Because of the oversampling, a relatively simple LPF is used following the D/A conversion process. (Very elaborate filtering of the audio signals would be required were it not for the preliminary filtering provided by the oversampling process.) Note that some CD players use an 18-bit digital filter to provide 8-times oversampling. However, this is not the usual case for either CD or CDV players.

2.3.5 Stereo Audio Amplifiers

A CD player includes some form of audio amplification after the S/H and LPF circuits. Generally, the audio amplifiers are in IC form, and they raise the left- and right-channel audio to a level of about 2 V. In some cases (particularly the early-model CD players), the audio output level is fixed. However, most late-model CD players have an adjustment (volume control) to set the audio output level. Also most CD players have some form of stereo headphone output (usually adjustable).

2.4 CD AND CDV OPTICAL PICKUPS (READOUTS)

The following paragraphs describe additional optical pickup principles not covered in Chap. 1. As discussed, digital information contained on the CD or CDV is retrieved by means of a laser which generates a light beam to detect the track of pits and flats on the disc. The laser beam is applied to, and reflected from, the track through an optical pickup or readout (a series of lenses, prisms, mirrors, etc.).

Remember that there are several types of optical readouts for both CD and CDV players currently in use, and new systems are being developed. However, as discussed in Chap. 1, the optics are part of a pickup or readout assembly that must be replaced as a package (in most players). We describe some typical replacement procedures in Chap. 7.

2.4.1 Focus and Tracking

In addition to recovering the audio and/or video information, the optics also provide for laser-beam focus and tracking of the pits and flats. Present-day players provide for focus and tracking by means of focus and tracking coils which move the objective lens in relation to the CD (Chap. 1). The coils are driven by error signals that originate from the optical-system photodiodes.

Older CDV players (such as the many Laservision players still in use) provide tracking by means of movable mirrors. Such players provide both radial and tangental tracking (in addition to focus correction), as discussed in Sec. 2.4.4.

2.4.2 CD Optics

Figure 2.7 shows the components for typical slide-type optics (the most commonly used type of CD optics). When the laser diode is activated, light is emitted

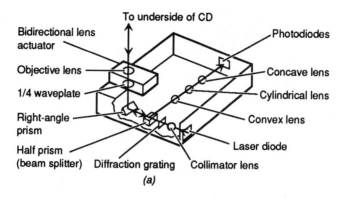

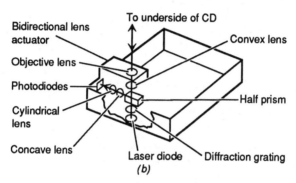

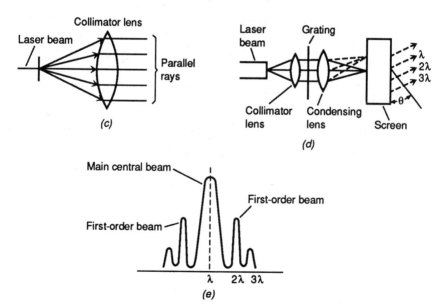

FIGURE 2.7 Components for typical slide-type optics.

and passes through a *collimator lens*. The function of this lens is to take the light rays of the laser beam and make them parallel, as shown in Fig. 2.7c. (In this case, the laser is considered as a point source.) The parallel rays from the collimator lens enter the *diffraction grating*.

A diffraction grating is a screen that contains a very large number of very narrow slits, uniformly spaced at distances only a few times the wavelength of the laser beam. When a beam of light passes through narrow slits, the exiting beam not only continues in a straight line but also diffracts at different angles, as shown in Fig. 2.7d. (If you focus the exiting beam on a screen, you see a row of bars or dots with the center dot or bar being the brightest, and the dots on both side decreasing in intensity.)

In the optics of a CD player, we are concerned with the main center beam (central image) and the two beams (first-order image) that are to the immediate right and left of the center beam. The center beam is used for reproduction of the audio, tracking, and focusing, while the two first-order beams are used for radial tracking only.

The three beams from the defraction grating go through a *beam splitter* (also called a *half prism*). The beam splitter is like a one-way mirror or window mounted at 45°. When the three beams from the laser approach the beam splitter, one side looks like a window. To the beams returning from the CD, the splitter looks like a mirror. The beams from the beam splitter are reflected (at a 90° angle) to the optics in the bidirectional lens actuator by a *right-angle prism*.

The next device in the path from the laser diode to CD is the *¼ waveplate*. A ¼ waveplate is an optical device that produces and detects circularly polarized light. (This is the physics definition of the ¼ waveplate.) For our purposes, consider the ¼ waveplate as a device that changes some of the properties of the beam to distinguish which beam is being reflected from the CD and which beam is coming from the laser diode.

The beams reflect, or do not reflect, depending on the absence or presence of a pit. The reflected beam returns to the beam splitter (half prism) via the objective lens, ¼ waveplate, and right-angle prism. At the beam splitter, the beams reflect from the mirrored surface and are applied (at another 90° angle) to the photodiodes through a series of lenses (typically convex, cylindrical, and concave).

Note that Fig. 2.7 shows two optical-pickup configurations. In Fig. 2.7a, the right-angle prism is required to reflect the beam up to the CD since the laser diode is mounted on the side of the pickup assembly. In Fig. 2.7b, the right-angle prism and collimator lens are omitted since the laser diode is mounted directly in line with the objective lens. The configuration in Fig. 2.7b also has a different placement for the convex, concave, and cylindrical lenses.

2.4.3 CDV Optics

Figure 2.8 shows the components for the optics of a CDV player. Note that these optics are used to play both CD and CDV discs of all types (including the newer Gold CD, older Laservision, and so on).

Laser Beam. The laser diode emits a light beam which first passes through the grating to split the beam into a primary and two secondary beams. The center

ENCODING, DECODING, AND OPTICAL READOUT

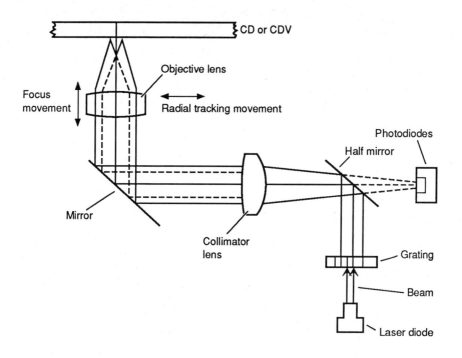

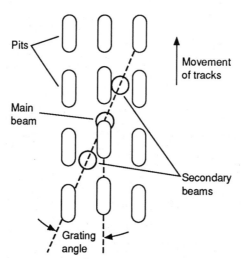

FIGURE 2.8 Components for typical CDV optics.

beam, which is the brightest of the three beams, is used to focus the beam of the reflective surface of the disc and to read the tracks on the disc. The secondary beams (one at each side of the main beam) are used strictly for radial tracking (and are also known as *radial beams*).

Part of the light (some light passes through the mirror and is lost) from the three beams (light bundle) is reflected from the half mirror toward the collimator lens, which forms a column with the three beams parallel to each other. The light bundle exits the collimator lens toward the mirror and is directed up through the objective lens, which focuses the laser beam onto the reflective surface of the disc. The reflected light bundle returns through the objective lens, the mirror, and the collimator lens. About half of the reflected light bundle passes straight through the half mirror to strike the photodiodes.

The reflected light from the main beam is intensity-modulated by the pits in the track that moves across the beam at a high velocity as the disc spins (Fig. 2.8*b*). The secondary beams strike the disc in between tracks, as shown. The angle between the secondary beams and the main beam is known as the *grating angle* and is set by the *grating angle adjustment*.

Focus. Figure 2.9 shows the photodiode arrangement and focus principle. There is some wobbling as the disc spins over the laser beam since a disc is never perfectly flat. Therefore, there is a need to constantly adjust the focus of the laser beam on the reflective surface as the disc is spinning. An elliptical shape of the beam occurs if the disc is too close to or too far from the objective lens. For example, if the disc is too far from the objective lens, photodiodes D_1 and D_2 receive more light than D_3 and D_4. The opposite occurs if the disc moves too close to the objective lens.

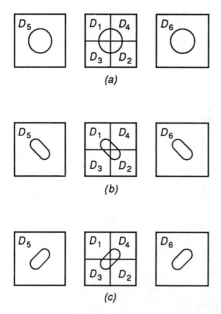

Photodiode Operation. All of the photodiodes operate in the photoconductive (reverse biased) mode. The arrangement of the photodiodes allows the three beams to fall, as shown in Fig. 2.9. The center beam falls on the four center photodiodes D_1 through D_4, and the secondary beams fall on the two adjacent photodiodes D_5 and D_6. The elliptical shape falling on D_5 and D_6 has no effect on the operation of the player.

If the beam is perfectly focused on the reflective surface of the disc, the center beam falls equally on each of the four center photodiodes D_1 through D_4. Therefore, the sum current of photodiodes D_1 and D_2 is equal to the sum current of photodiodes D_3 and D_4.

If the beam is out of focus, the shape of the beam falling on the photodiodes becomes elliptical and the total current of one pair of photodiodes is greater than the other pair. The difference in the photodiode pairs creates the focus error (FE) signal which is used by the focus servo to make the necessary focus corrections.

FIGURE 2.9 Typical photodiode arrangement and focus principles. (*a*) Disc perfectly focused; (*b*) disc too far from lens; (*c*) disc too near lens.

Radial Tracking. Figure 2.10 shows the method used to maintain proper radial tracking. When the main beam is directly centered over the track, the radial beams fall just outside the track. If the beams move either to the right or left of center, the intensity of one reflected beam is greater than the other. For example, when there is a shift to the right of the track, the reflection of radial beam R is greater than the reflection from radial beam L. The reflection is less from radial beam L because much of the beam is falling on the track which contains pits.

Since much of the light falling on a pit is diffused, less light intensity reaches the corresponding photodiode D_5. Therefore, the current from D_5 is less than the current from D_6. Photodiodes D_5 and D_6 are equally illuminated when the beam is perfectly centered on the track. The current of D_5 is therefore equal to the cur-

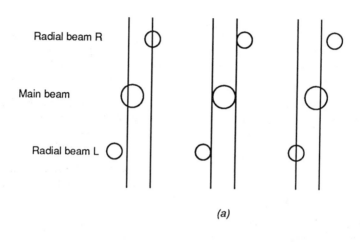

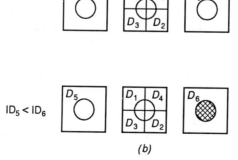

FIGURE 2.10 Typical photodiode diode arrangement and radial-error principles.

rent of D_6. The radial error (RE) signal is created by the difference in current between D_5 and D_6.

2.4.4 Laservision (LV) Optics

Figure 2.11 shows the components for the optics of a Laservision, or LV, player. Note that the LV player uses a helium-neon (H-N) laser, as well as fixed and movable mirrors. This system has generally been replaced by the CDV optics discussed in Sec. 2.4.3. However, there are still many LV players in use. Before we discuss operation of the LV optical system, let us review the characteristics of light and light polarization (for those who are interested).

Light Characteristics. Light is electromechanical energy with wavelengths that are visible to the human eye. For our purposes, think of a light beam as a high-frequency sine wave passing through space. (The red light emitted by the H-N laser has a wave length of 6328 Å and is generally visible to the human eye. The beam produced by a solid-state laser is generally not visible and has a wave length of about 7900 Å.)

Generally, a light wave changes *direction of polarization* at random and is referred to as *nonpolarized* light. Various light polarizations are shown in Fig. 2.11b. If the direction of polarization is constant, the light is said to be *linearly polarized*, and polarization can be vertical, horizontal, or circular.

The LV player uses optical components which are affected by the polarity of light, as are polarized sunglasses. The special lenses of polarized sunglasses do not pass light that is polarized horizontally. Since most light from glare and reflection is horizontally polarized, the light does not pass through the lenses.

LV Optics Operation. The red light beam from the laser shown in Fig. 2.11a is vertically polarized. The first optical component encountered by the vertical light beam is the *first fixed mirror*, which reflects the beam around a corner to the *second fixed mirror*. In turn, the second fixed mirror reflects the beam around another corner and onto the *grating*. Note that the fixed mirrors serve only to fold the beam around so that the optics can be housed in a compact space.

The grating (a piece of optical glass with several fine etched horizontal lines) divides the main beam into three beams: a center or main beam and two secondary beams (above and below the main beam). The secondary beams are less bright than the main beam. (Actually, the grating creates more than three beams, but other beams are of such reduced brightness that the additional beam can be ignored.)

The center, or main, beam is used to read the tracks on the videodisc, while the secondary beams are used for tracking. The three beams are collectively referred to as a *light bundle*. For our purposes, we consider the light bundle as a single beam.

The light bundle from the grating is applied to a *prism* through a *diverging lens*, which focuses the beam to the correct size to completely fill the aperture of the *objective lens*. The prism bends the light in a direction determined by the polarity of the light beam. The vertically polarized light from the laser is bent in the opposite direction to horizontally polarized light.

As the beam exits the end of the prism, the beam passes through the ¼

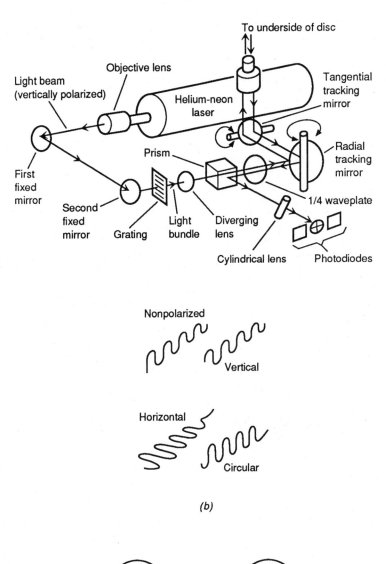

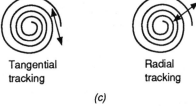

(b)

(c)

FIGURE 2.11 Components for typical Laservision optics with helium-neon laser.

waveplate, which changes the beam into a circularly polarized beam. The circular beam is then reflected from the *tracking* and *tangential mirrors* into the objective lens.

As shown in Fig. 2.11c, *radial tracking* refers to keeping the light beam centered on the tracks. Without some form of radial tracking, the light beam can drift between the tracks, resulting in a lost picture. Radial movement of the beam is always perpendicular to the tracks. *Tangential tracking* refers to keeping the beam in line with the track. This direction of movement is necessary to compensate for momentary speed errors as the track passes over the beam. Both radial and tangential tracking are accomplished by the two movable mirrors.

As in the case of CD and CDV optical systems, the diodes generate an RE voltage if the beam tends to drift off the track. The RE voltage is applied to the radial-tracking mirror through a radial-tracking servo. The error voltage moves the mirror to bring the beam back to the center of the track.

In the LV player, signal-processing electronics generate a tangential-error voltage which is proportional to the momentary speed errors in tracking. (The tangential-error voltage is in addition to a turntable speed-error voltage.) The tangential-error voltage is applied to the tangential-tracking mirror through a separate tangential-tracking servo. The error voltage moves the mirror to move the beam as necessary to compensate for the momentary speed error (often caused by slightly elliptical tracks or an offset center hole in the disc).

Operations of the objective lens, cylindrical lens, and photodiodes for the LV player (Fig. 2.11a) are essentially the same as for corresponding components in a CD or CDV player (Figs. 2.7 and 2.8).

2.5 CDV TRACK FORMATS

The following paragraphs describe additional CDV tracking principles not covered in Chap. 1. This includes converting movie film to CDV videodiscs, CDV special operating modes, CDV track-jumping formats, and CDV search modes.

2.5.1 Converting Movie Film to LV Videodiscs

Movie cameras use a frame rate of 24 Hz, while TV uses 30 Hz. To overcome this problem, movie film is converted to TV fields according to the format in Fig. 2.12a when recorded on videodiscs. The film frames are alternately scanned for three video fields and then two video fields, as shown. So five video frames are consumed while scanning four film frames.

Picture numbers are added only to the *first* video field per film frame. Since only four picture numbers are used for each five video frames, the picture numbers reach only about 43,200 ($4/5 \times 54,000$) at the end of a 30-min videodisc.

2.5.2 Videodisc Special Operating Modes

Videodisc players are capable of many special operating modes, such as reverse play, fast forward, still picture, and slow motion (all under control of a micropro-

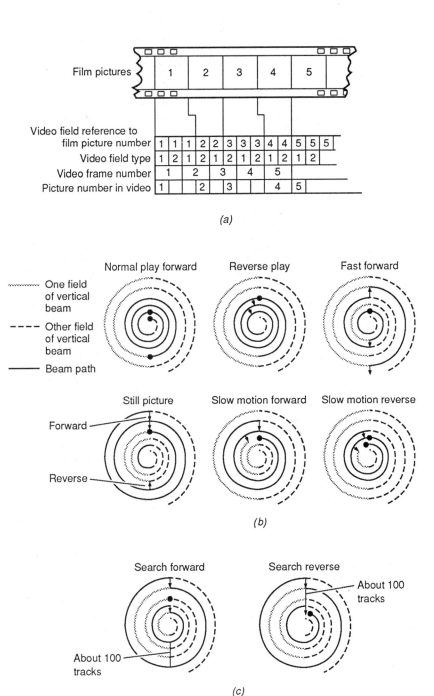

FIGURE 2.12 Converting movie film to videodiscs.

cessor, as discussed in Chap. 6). These modes are possible because the beam can be made to "jump" from track to track at appropriate times by controlling the radial-tracking servo.

To prevent the jump from being seen on the TV screen, all track jumping takes place only during the TV vertical sync interval. Since two vertical sync intervals occur for each revolution of the disc, a maximum of two jumps per revolution can take place. Note that this condition of two sync intervals per revolution is found only on CAV videodiscs. As a result, the special operating modes are not available on CLV videodiscs.

2.5.3 Track-Jumping Formats

Figure 2.12*b* shows the track-jumping formats required to create the various modes of operation. Each spiral drawing represents a few tracks of the videodisc. The left half (solid line) is one field of the TV vertical scan, and the right half (dotted line) is the other field of the vertical scan. The dark line represents the path of the beam. The path begins at the heavy dot in each illustration of Fig. 2.12*b*.

In the *normal play mode* no track jumping takes place. The beam follows the spiral track from the inside to the outside. *Reverse play* requires the beam to jump back one track after each field. At the end of each revolution, the beam is on track inside of where the beam is started. The result is *normal speed reverse play*.

A *still picture* occurs when one track is continuously repeated. One track is read and then the beam jumps back one track and repeats. The result is the same whether the forward or reverse Still-Picture button is pressed on the front panel. However, once the still-picture mode is in effect, additional actuations of the Still-Picture button makes the picture move forward or backward.

When the Still-Picture Reverse button is pressed, the beam is made to jump inward on track (shown as reverse in Fig. 2.12*b*). However, this jump takes place on the *field opposite* that of the previous jump. (If this were not so, the beam would have to jump two tracks to get the original track.) Pressing the Still-Picture Forward button allows the beam to continue on to the next track. After one revolution, the beam simply begins repeating the next track.

Slow motion is a series of still pictures, each one lasting for a certain number of revolutions. The slow-motion control determines how long each still picture remains on until the beam continues to the next track and creates the next still picture. The effect is the same as rapid manual actuations of the Still-Picture Forward button. *Slow-motion reverse* uses the same principles, except the still pictures advance in the reverse direction.

Fast forward is created by jumping forward one track after each field. After one revolution, the beam is three tracks further out than it was. As a result, the fast-forward speed is actually 3 times the normal forward-play speed.

2.5.4 Search Modes

Figure 2.12*c* shows the beam action during videodisc search modes. When search is activated, the optical pickup (slide assembly) begins to move rapidly across the bottom of the videodisc. Sections of the picture are momentarily displayed during the search process.

During *search forward*, the beam displays a series of pictures and then skips several hundred tracks, displays another series of pictures, and so on. These pictures are displayed very rapidly because the optical pickup takes only about 25 s to scan the entire videodisc. The *search-reverse* action is the same except that the beam moves inward when skipping the tracks.

CHAPTER 3
USER CONTROLS, OPERATING PROCEDURES, AND INSTALLATION

Although CD players are not difficult to operate or install, the basic procedures are quite different from those of a typical LP player. The same is true of CDV players. This chapter describes the basic user controls and operating procedures for CD and CDV players. We also discuss typical installation procedures and precautions. A careful study of these procedures will help you understand the operation of the player circuits (which are discussed in Chaps. 5 and 6) and the mechanical assemblies (Chap. 7).

Remember that you must study the operating controls and indicators of any CD or CDV player you are servicing. This book describes "typical" controls and indicators, but there are subtle differences in operation you must consider. For example, the disc compartment of some players must be closed manually, while most others are automatic. In some players, the compartment closes automatically when the Play button or key is pressed. In other players, you must push the Open/Close key (to close the compartment) before you press Play. There is nothing more frustrating than troubleshooting a failure symptom when the player is supposed to work that way.

3.1 OPERATIONAL SAFETY CHECKS

It is assumed that you are familiar with basic operational safety checks for consumer electronic equipment (such as a *leakage current test*). Virtually all service manuals describe the necessary safety procedures in full detail, so we will not repeat the tests here. However, we recommend that you follow all of the procedures (for your own sake and for the sake of the customer).

As an important example, if there is current leakage, indicating that metal parts of the player are in electrical contact with one side of the power line, this can result in damage to the player or possible shock to anyone touching the exposed metal parts. So, before you release a player to the customer (after service or when the player is first sold), check for any leakage current as described in the service manual.

In addition to operating checks, there are certain safety precautions to be performed during service. These are discussed in Chap. 4.

3.2 TRANSIT OR SHIPPING RESTRAINTS

Many CD and CDV players have a transit or shipping screw to hold the optical pickup in place when the player is moved or shipped. Without such a restraint, the optical assembly can move back and forth, causing possible damage to the delicate optics. Typically, the transit screw is accessible from the bottom of the player.

Always remove or loosen the transit screw *before* using the player. In some players, the screw is captive and cannot be removed (the screw is to be loosened only). In others, the screw is supposed to be left in place (after it is loosened) but can be removed if you keep turning the screw. In other players, the screw is supposed to be removed. Be sure not to lose the screw. (It is the customer's job to lose it.)

Make sure the screw is properly installed *when transporting or shipping the player*. Typically, the screw can be installed only when the optical pickup is in one position (the at-rest or secured position). For example, you must turn on the power, make sure there is no disc installed, and then shut the tray, drawer, lid, or door electrically. With the optical pickup in the secured position, turn off the power and install the screw.

The transit screw brings up an obvious problem. When the customer brings in a CD player to the shop for service, he or she will probably forget the screw or not tighten it, possibly damaging the optics. The opposite occurs if you tighten the screw upon returning the player to a customer. When they get the player home, they will promptly call you with a complaint that the player "no longer works after you fixed it." You must patiently explain these facts to the customer. Good luck.

3.3 EXTERNAL CONNECTIONS

External connections between a CD player and stereo system are usually made at the back of the player, as shown in Fig. 3.1. Connections between a CDV player and the TV receiver or monitor are also made at the back, as shown in Fig. 3.2.

Note that the stereo audio output of most CDV players can also be connected to a stereo system if desired. On certain advanced CDV players (such as the Philips CDV488), the digital audio output (digital bit stream) is also available at the back connectors. These outputs are for transmission over a fiber optics cable and/or digital coaxial cable to an advanced CD player [so that you get the benefits of the CD player's superior digital-to-analog (D/A) decoding and reproduction] before the audio is applied to the stereo system. Note that most CDV players *do not* provide this digital bit stream output.

3.3.1 CD Electrical Connections

Typically, a pin cord is supplied with the player (or possibly the amplifier) for connection between the player L and R stereo outputs and the corresponding inputs on the amplifier. Although the connections are very simple, certain precautions must be observed (for all players).

Never connect a CD player to the Phono input of an amplifier. Instead, use the

CONTROLS, OPERATION, AND INSTALLATION

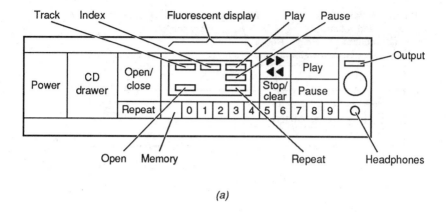

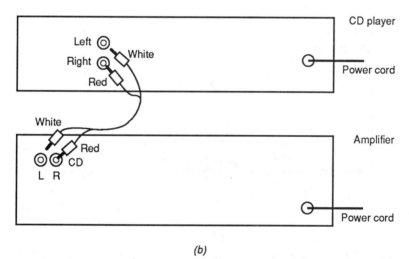

FIGURE 3.1 Operating controls, indicators, and connections for composite CD player.

CD Aux or possibly the Tape Play inputs. The CD player output is about 2 V, which can damage the amplifier and/or speakers and will overdrive the amplifier.

Even on those CD players where the output is adjustable, the player impedance (50 k) is best matched to the CD/Aux/Tape inputs. As is the case with any audio component, always switch off the power before making or breaking connections between the player and amplifier.

Before connecting the player to a power source, check that the operating voltage of the player is identical to the voltage of the local power supply. Unlike most home electronic components (TV, VCR, etc.), many CD players are designed for worldwide use and can be operated at 120, 220, and 240 V. The operating voltage

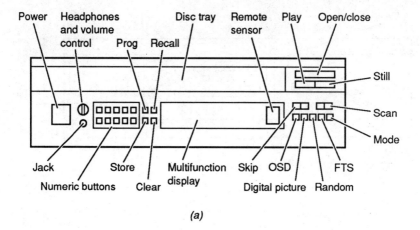

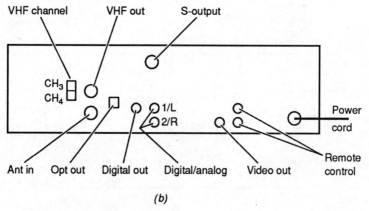

FIGURE 3.2 Operating controls, indicators, and connections for composite CDV player.

is selected by special connections and/or switch settings, so check the service or operating literature.

3.3.2 CDV Electrical Connections

As shown in Fig. 3.2, the rear-panel connectors for a CDV are similar to those of a VCR. (If you are not familiar with VCRs, read *Lenk's Video Handbook*, McGraw-Hill, 1991, immediately).

CONTROLS, OPERATION, AND INSTALLATION   3.5

If the CDV is used with a standard TV (not a monitor), the composite RF signal from the CDV is applied to the TV through the VHF Out connector. The output from VHF Out is either (1) a TV broadcast signal or (2) the CDV video and audio signals converted to a VHF channel 3 or 4 signal by the built-in RF converter, as selected by the VHF Channel switch. A VHF antenna (or cable input) is connected through the ANT In connector.

If the CDV is used with a monitor-type TV, the video from the CDV is applied directly to the monitor through the Video Out connector, thus bypassing the RF converter. Likewise, the monitor can receive audio directly through the 1/L and 2/R Digital/Analog connectors.

If the TV or monitor is capable of super-VHS (S-VHS) *operation*, the S-VHS video is applied to the monitor through the S-Output connector (as is the case for a VCR capable of super-VHS operation).

The CDV audio can also be applied to a stereo system or amplifier and speaker combination through the 1/L and 2/R Digital/Analog connectors.

The Remote Control connectors are used only when the player is controlled from other components through a remote-control system (the Philips RC-5 system in this case). The hand-held remote control is normally used with the front-panel remote sensor, as discussed in Sec. 3.6.

If the CDV player is to be used with an advanced CD player (such as the Philips LHH1000) to take advantage of the superior decoding and reproduction (Sec. 3.3), both the Opt Out and Digital Out connectors are used.

The Opt Out connector outputs the CD, CDV, and CDV-LD audio signal in digital format for transmission over an optical-fiber cable. The Digital Out connector provides a similar output in electrical form. As discussed in Sec. 3.3, these digital outputs are not found on most CDV players.

3.4 GENERAL OPERATING AND INSTALLATION NOTES

The following precautions and tips should be followed when operating any CD or CDV player. Chapter 4 also includes a series of precautions and tips that apply primarily to service and handling.

3.4.1 CD Operating and Installation Notes

Again, check that the operating voltage of a CD player is identical with the voltage of the local power supply. As discussed, some CD players can be operated at 120, 220, and 240 V, while other players can be operated at only one specific line voltage. Usually, there are special connections, or switch settings, required for dual-voltage players. Always check the service and/or operating literature for such connections or settings.

Figure 3.3 shows the power-line connections for a CD player with a rear-panel voltage selector switch. Note that the correct tap on the player power transformer is selected by the switch.

When the CD player is not in use, turn off the power. This conserves energy and extends the life of the player. Unplug the player from the power output if the

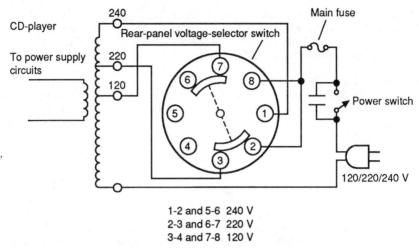

FIGURE 3.3 Power-line connections for a CD player with rear-panel voltage-selector switch.

player is not to be used for an extended period (and is not part of a system, such as a modular component that receives power from an amplifier).

Any type of player (CD or CDV) that uses a laser beam produces strong magnetic fields. Do not place video or audio cassettes on or near the player.

Do not install the player near heat sources such as radiators or air ducts or in a place subject to direct sunlight, excessive dust, mechanical vibration, or shock.

Good air ventilation is essential to prevent internal heat buildup in the player. Place the player where adequate air circulation is assured. Pay particular attention to the top ventilation holes (which should never be obstructed).

Do not place the player on a soft surface, such as a rug, which might block the bottom ventilation holes.

If the player is brought directly from a cold to a warm place, or is placed in a very damp room, *moisture may condense on the optical-pickup lenses*. The player will not operate properly (if at all) should the lenses become fogged. In such a case, remove the CD and leave the player turned on for about an hour to evaporate the moisture.

Although CDs are not delicate and should last forever, *CDs should be handled with some care*. For example, always handle the CD by the edges. Try to avoid touching the rainbow-colored surface. Smudges or other dirt on this surface can reflect the laser beam and cause dropouts or other undesired performance.

Do not stick paper or tape on the labeled surface. While such paper or tape will not hurt the CD, it can get caught in the drive mechanism (particularly the clamp). Do not expose the CD to direct sunlight or heat sources, such as hot-air ducts, or leave a CD in a car parked in direct sunlight (where there can be considerable rise in temperature).

Before playing, clean the CD with a soft cloth. Many CD players include a cleaning cloth (usually packaged along with the pin cord and operating instruction manual). After playing, store the CD in its case. *Never use solvents* such as benzine, thinner, commercially available cleaners, or antistatic spray intended for LP records. Any of these can eat through the sealant and destroy the CD. However, there are cleaners specifically designed for CDs (Chap. 4).

Some CD players cause interference to radio and TV reception. Usually, the problem can be eliminated if the player is moved a few feet away from the radio or TV.

Setting CD Player Volume Controls. Probably the most important operating precaution for those not familiar with CD players involves setting the volume controls. The dynamic range of the CD player is much greater (90 dB or greater) than any LP record player, and the peaks are recorded with hifi. Also, you get a much greater signal-to-noise (S/N) ratio (also 90 dB or greater) with a CD player. Background noise is practically eliminated in most CD players.

If you turn up the volume, either on the player or the stereo amplifier, while listening to a portion of the CD where no audio signals (or very low-level signals) are recorded, the speakers may be damaged when that portion of the CD with peak signals is played. Likewise, if you are listening with stereo headphones connected to the front panel of the player and set the player volume in an attempt to hear background noise, you can drill a hole from ear to ear when you get to the audio peaks.

3.4.2 CDV Operating and Installation Notes

Most of the operating notes for CD players (Sec. 3.4.1) also apply to CDV players. However, the installation procedures are quite different. Installation of a CDV involves connecting the player to an antenna or cable, a monitor TV, the power source, and possibly to a stereo and/or CD player. Although the following notes apply to a specific CDV (the Philips CDV488), the connections are typical.

Connecting the VHF Antenna. If the antenna (or cable outlet) uses a 75-Ω coaxial cable, disconnect this cable from the TV set and reconnect the cable to the Ant In connector shown in Fig. 3.2*b*. (If the antenna is of the 300-Ω ribbon type, an adapter is required between the antenna cable and the Ant In connector.) Next, connect the player VHF Out connector to the TV set antenna input, in place of the antenna (or cable outlet) coaxial cable. Often, a coaxial cable with F-type connectors on both ends is supplied with the player for this purpose. Again, use an adapter if necessary to match the 75-Ω output of the player to the TV set.

Setting the Channel Selector. Set the player VHF Channel switch to Channel 3 or 4, *whichever is not used* for commercial TV broadcasts in the area. Then set the TV to the *same channel*.

Connecting Power. Connect the ac power cord to a wall outlet (120 V, 60 Hz).

UHF/VHF Installations. If there is a separate UHF antenna and cable, leave the UHF antenna connected to the TV set. Where a combination VHF/UHF antenna is used, separate the VHF and UHF signals using a signal separator. Connect the VHF lead-in to the player and the UHF lead-in to the TV set. Most combination antennas have a built-in signal separator, but you may need a separator for some installations.

Control of the VHF Signal. With the connections described thus far, VHF signals from the antenna and player to the TV set are switched by the player.

Installation with a VCR. If you are using a VCR (with a built-in TV tuner), connect the output cable from the VCR VHF connector to the player Ant In connector. Then connect the player VHF Out connector to the TV set antenna input, in place of the VCR cable. Remember that with this installation, *both* the VCR and CDV must be switched off (or as directed in the VCR and CDV operating instructions) before the TV set is connected to the antenna (or cable outlet).

Caution. Connections between the player VHF Out connector and the antenna terminals of a TV set should be made only as described in this section (or as specified in the operating instructions for the player). Failure to do so may result in operation that violates FCC regulations regarding the use and operation of RF devices. *You may broadcast CDV programs to the entire neighborhood.* Never connect the output of the player (VHF Out) to an antenna or make simultaneous (parallel) antenna and player connections at the antenna terminals of the TV.

Installation with a Monitor TV. If the player is used with a monitor TV, connect the player Video Out terminal to the corresponding Video In terminal on the TV. Then connect the Digital/Analog 1/L and 1/R terminals to the corresponding audio-input terminals on the TV. The video connection requires a coaxial cable, while the audio is usually made through a pin cord. If the monitor or TV is capable of S-VHS (super-VHS) operation, connect the player S-Output connector to the corresponding S-VHS input on the TV. (This requires a special S-VHS cable.)

Installation with a Stereo System and TV. If the player is used with a stereo system, connect the Digital/Analog 1/L and 2/R terminals to the corresponding audio-input terminals on the stereo (usually at the Auxiliary input or the tuner-input jacks on the stereo amplifier or receiver). For best results, place the TV set (on which the player video is displayed) *between* the stereo speakers. Turn down the TV volume control to minimum to eliminate any audio from the TV speakers. Note that on a few very old CDV players, it is necessary to select stereo operation.

Installation with a CD Player. If the CDV player is to be used with a CD player (or any system where the digital bit stream from the player optical system is used directly without D/A processing) make the connections at the Opt Out and/or Digital Out connectors. The Opt Out connector requires a fiber optics cable, while the Digital Out connector uses a coaxial cable.

Antenna Signal Level. Make certain that the signal level from the antenna is adequate. Inside any CDV player, the antenna signal is distributed to the TV set through some form of switching network (usually a coaxial switch). If the antenna output is very low, signal loss caused by the distribution may result in deterioration of the off-air TV picture (when the CDV is not in use). The antenna-switching circuits of properly functioning CDVs are well designed and introduce a minimum of loss. However, it is possible to have trouble with a "borderline" TV antenna. Needless to say, the customer will blame the CDV.

Fine Tuning. Check that the TV fine tuning is properly adjusted for the *inactive channel.* The player output signal uses the inactive channel of the TV, but since this channel is not ordinarily used, the fine tuning may not be precisely adjusted. Play a disc of known quality and adjust the TV fine tuning for best reception. The

VHF output of most players is not adjustable (cannot be fine tuned). However, this should be checked on older CDV players.

Battery Replacement. Some CDV players (such as the Philips CDV488) require a battery to maintain power on the memory functions. This keeps customer programming information in memory when the player is disconnected from power. Usually, a lithium battery is used and need not be replaced frequently (usually they last for years). However, always check the battery (if any) when you encounter a "this player won't stay programmed" trouble symptom. Also check the battery when the player is first installed.

Demonstration. If you have the job of demonstrating the use of a CDV player to customers, go over the operating instructions in the user manual with them in detail. Although operation of a CDV is simple to an electronic wizard such as yourself, it may not be so to the general public, since CDV has many capabilities beyond those of a record player, CD, stereo system, or a TV set.

3.4.3 After-Installation Checkout (Preliminary Troubleshooting)

Before concluding that a CDV player is out of order, run through the following notes and make sure that there is not a simple remedy for the problem. When a CDV player is first placed in operation, the vast majority of performance problems are setup errors, bad connections to other equipment, or malfunctions of other equipment (TV, stereo, antenna, etc.). So, before you get into the detailed service notes and troubleshooting described in Chap. 9, where we discuss specific troubles related to the major functional sections of CDVs, let us review some simple, obvious steps to be performed (before you start actual troubleshooting).

If the disc tray (or lid on some players) does not open, check that the power cord is plugged in and that the Power button or key is pressed. If these problems check out, try pressing the Open/Close key.

If the disc does not appear to rotate (if you cannot hear the disc motor running), check that the power cord is plugged in and that the Power button is pressed. Also check that the disc tray is fully in (to close interlock contacts).

If the disc stops rotating soon after running, it is possible that you are playing an unrecorded side of the disc. Try turning the disc over.

If the disc rotates (motor running) but there is no picture, check the following:
Is the TV set turned on? Turn it on.
Are the connections between player and TV set (and/or stereo) correct (Fig. 3.2b)?
Is the TV set turned to channel 3 or 4? Set the TV to the channel (3 or 4) that is *not used* for TV broadcasting in the area.
Is the player set to the *same channel* as the TV?
Is the player in a pause mode? Press the Play button.

If the picture quality is bad and you have checked the connections between player and TV set and have checked that both are on the same channel, check the *fine tuning* on the TV set. If the picture quality is still bad, try a different disc (a known good disc).

If the TV no longer receives other channels after being connected to the player, make sure that the antenna cable is connected to the player (antenna cable to Ant In connector on player). Then check that the player is in the off condition (Power switch off, etc.). Remember, to view regular TV broadcasts, the player must be turned off (using whatever buttons, controls, or levers are required).

If only a particular part of a disc does not produce a good picture, the disc is probably damaged. Try pressing the Scan (or Rapid Access or Visual Search) buttons to skip over the damaged portion.

If the hand-held remote control does not work but the player can be operated by the front-panel controls, try replacing the remote-control batteries. Also check for obstacles between the player and the remote control and that the remote control is aimed directly at the player (preferably at the remote sensor, Fig. 3.2*a*).

3.5 BASIC CD PLAYER USER CONTROLS

Figure 3.1*a* shows the operating controls and indicators for our CD player (a composite of many players). Compare the following with the controls and indicators of the player you are servicing (which, of course, will be different). Press Power to turn the player on or off. Press Open/Close to alternately open or close the CD drawer. When the drawer opens, a red Open indicator turns on. Pressing Open/Close while a CD is playing stops the mechanism and opens the drawer, the repeat function is canceled (if on), and program memory is erased.

Display Track number: When the CD drawer is closed with the Open/Close button, when the CD has played to the end and stopped, or when the Stop/Clear button is pressed, the total number of tracks on the CD is displayed (on the fluorescent display). At all other times, the track number being played is displayed.

Display Index number: If index decoding exists on the CD, the index number is displayed while the CD is played.

Display time indicator: When the number indicating the total number of tracks is displayed, the total playing time is also shown. At other times, the elapsed playing time is displayed for the track that is being scanned.

Display Play indicator: Play turns on during normal play.

Display Repeat indicator: Repeat turns on when the repeat function is active.

Display Pause indicator: Pause turns on when the pause function is active.

Display Memory indicator: Memory turns on when a program sequence is in memory.

Display Open indicator: Open turns on when the CD drawer is opening.

The button with double arrows pointing to the right scans the CD in a forward direction. When first pressed, the audio material is heard at about 3 times normal speed. If the button is held, the player begins sampling short segments. Release the button to resume normal play. The button may also be pressed with the Play button to skip to the beginning of the next track.

Pressing *the button with the double arrows pointing* to the left scans the CD in

CONTROLS, OPERATION, AND INSTALLATION   3.11

a reverse direction. The button may also be pressed with the Play button to restart the selection (being played) at the beginning.

The Stop/Clear button stops playback and clears any program from memory.

The Play button initiates play. It may also be used to close the CD drawer and begin play.

Press Pause to interrupt playback. Press Play to resume playback. The Pause button may be used to close the CD drawer. The *directory of selections* (also called the *CD directory, disc directory*, or simply *directory*) is read, and the player pauses at track 1.

The Output control adjusts the output level to *both* the headphone jacks and to the rear-panel output jacks. (This is not always true on all CD players.) Make certain to observe the note in Sec. 3.4.1 regarding the Output control. It might save your ears.

The 0 through 9 buttons are used to select a track or index number. After a track number is selected, the display flashes that number. Push Memory to enter the track into memory. Repeat for each track to be programmed.

Push Repeat to repeat the entire CD or the selections that have been programmed into memory.

Note that most of the front-panel controls have remote-control counterparts. Most operating sequences can be initiated or ended using the front-panel controls, the remote unit, or a combination of both (on most players).

3.5.1 Preliminary Operating Checks

Always make a few preliminary checks at the initial installation and before launching into a full troubleshooting routine. Make certain that the transit screw is removed or loosened *before* operating the player. Tighten the transit screw *only* when the CD player is to be moved.

If practical, check that the *customer's stereo system* is operating normally before you do any extensive service on the CD player.

Cleaning the objective lens should be a routine part of initial installation and servicing. A dirty objective lens can cause a variety of symptoms (intermittent or poor focus, skipping across the CD, erratic play, excessive dropouts, to name a few). These same symptoms can also be caused by a *defective CD*. Try a known-good CD first.

Do not replace the pickup assembly, or make any adjustments on the pickup, before checking for mechanical problems that can affect the pickup. For example, look for binding at any point on the pickup travel, indicating that the rails or guides are adjusted too tightly. At the other extreme, if you hear a mechanical "racheting" or "chattering" when the pickup is moved, the rails may be too loose. Mechanical adjustments are described in Chap. 7.

If the CD player operates manually but not by remote control, check the remote-unit batteries. Then try resetting the power circuits by pressing the front-panel Power button on and off.

If the left or right speaker is dead when the CD player is used in a system, try playing another component connected to the audio system line (tuner, cassette deck, etc.). If operation is normal with the other component, suspect the CD player. Also check for a loose cable between the CD player and other components. Then temporarily reverse the left and right speaker leads. If the same speaker remains dead, the speaker is at fault.

If the left or right speaker is dead when the CD player is used in a nonsystem configuration, temporarily reverse the left and right cable connectors at the amplifier input from the CD player. If the same speaker remains dead, the amplifier (or speaker) is probably at fault. If the other speaker goes dead, suspect the CD player or the player output cable.

If there is no sound from either speaker, check for the following. Both speaker-selector switches may be turned off. The wrong speaker terminals may have been selected (on amplifiers with two sets of speakers). The speaker cables may be disconnected. The CD player cables may be disconnected or improperly connected. In a nonsystem installation, the Output level may be too low (control too far counterclockwise) or the amplifier output selector may be set for the wrong source. Try the amplifier with a different input.

If the sound is distorted, the Output level may be too high, or the CD player output may be connected to the Phono input of the amplifier.

If there is hum or noise (only when the CD player is used), check for the following. The shield of the audio cable from the CD player may be broken, or the connector may not be firmly seated in the jacks. The CD player may be too close to the amplifier. The magnetic fields produced by the amplifier may induce hum into the player circuits (not likely, but possible).

If the player does not start, check the following. Make certain that there is a CD in the tray, that the CD is properly loaded (not upside down; the label should be up), and that the CD is firmly seated on the supports. Also check to see if the CD is very dirty, scratched, or warped. It is also possible that moisture has condensed on the CD player optical system (particularly on the objective lens).

If the sound cuts or repeats at some point, there may be a very dirty spot on the CD. Clean the CD with a soft cloth and mild detergent. (Do not use any commercial cleaners unless specifically recommended for CDs.) It is also possible that there is a scratch on the CD. Try skipping the point where the sound cuts or repeats.

3.6 BASIC CDV PLAYER USER CONTROLS

Figure 3.2a shows the operating controls and indicators for our CDV player. Figure 3.4 shows the controls for the hand-held remote unit. Compare the following with the controls and indicators of the player you are servicing (which, again, will be different).

On most CDV players, power must be applied before any operation, including remote operation, may be attempted. After the power is applied, the player goes into the start-up mode to check for the presence of a disc. Both CD and CDV on the display flashes until a disc is detected. The type of disc, CD or CDV, is displayed when a disc is detected. When a CD or CDV Single is detected, the display indicates the total number of audio tracks and the total time of the audio tracks on the disc.

The *disc tray* slides out when the Open/Close button on the player, or the Stop/Open button on the remote unit, is pressed.

When the Play button is pressed after placing a disc on the disc tray, the disc slides into the player and play starts. Pressing Play in the stop mode starts play;

CONTROLS, OPERATION, AND INSTALLATION 3.13

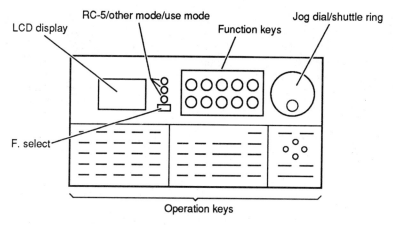

FIGURE 3.4 Controls for hand-held remote unit (universal learn remote).

pressing Play during play moves the play position to the beginning of the chapter or track being played and restarts play from there. Pressing Play can also start program play or favorite track selection (FTS) play.

Press Open/Close to open or close the disc tray. When Open/Close is pressed after placing a disc on the tray, the disc slides inside the player, the total playing time of the disc is displayed, and the player enters the stop mode. (Note that play is started as soon as the tray is closed. This depends on the information recorded on the disc.) The CD or CDV indicator blinks (or flashes) during the disc tray open or close operation.

When playing the video part of a CDV or CDV Single disc, pressing Still freezes the picture, but when playing the audio part of a CD or CDV Single disc, pressing Still places the player in pause. To release either the freeze or pause modes, press Still again.

Press Power to turn the player on or off.

Headphones can be connected to the Headphones jack for personal listening to the audio portion of the disc (the headphones signal is the same signal as that from the Digital/Analog jacks on the rear panel, Fig. 3.2b). The volume of the headphones is adjusted with the Headphones volume control.

Use the *numeric buttons* (1 through 10) when searching or programming chapters or tracks.

Press the Store button to enter a program for programmed play.

Use the Clear button in the following cases: to cancel repeat play, to interrupt search operation, to correct an entry made using the numeric buttons (holding Clear down clears the digit entry), to erase the last step in a program operation (holding Clear down clears the program entry operation), to cancel programmed play, to cancel random play, and to clear an FTS program.

The Prog button is used to program the desired chapters or tracks in a desired order (programmed play).

The *remote sensor* receives the signal transmitted from the hand-held remote unit.

Press one of the Skip buttons to skip to the beginning of a chapter or track. When the Skip button with double arrows pointing to the right is pressed, the *beginning of the next chapter or track* is detected. If the right Skip button is held, the chapter or track number is advanced continuously. When the Skip button with double arrows pointing to the left is pressed, the beginning of the *current chapter or track* is detected. When the left Skip button is held, the chapter or track number is reversed continuously. The Skip buttons are also used to skip between programs.

Press the On-Screen Display (OSD) button to switch the contents of the TV on-screen display and indicators. The first press of OSD shows the on-screen display; further presses switch the displayed contents. With some discs, the OSD button functions as an on-off switch for the on-screen display.

The Digital Picture button functions only when a CDV or the video part of a CDV Single is played. Press Digital Picture to select between a normal picture and a digital picture.

Press the Random button to start random play.

Press the FTS button to start automatic performance play. FTS is also used to enter the FTS program.

The Mode button functions only when a CDV Single is played. Press Mode to switch between the audio and video parts of a CDV Single disc.

When one of the Scan buttons is pressed and held during play, play is scanned forward or backward. The scanning speed varies in two steps (low speed for the first 2 s, then high).

3.6.1 CDV Player Remote-Unit Controls

The hand-held remote unit (Fig. 3.4) is a *universal or learn remote* unit capable of operating not only the CDV488 player but also a satellite receiver, a cable box, a TV receiver, an audio amplifier, and audio tape deck, an AM/FM tuner, and up to two VCRs. The following is a summary of the remote control functions.

Press the *Stop/Open/Close* key to stop play or to open or close the tray. During play, the player enters the stop mode with the first press of the key. The disc tray slides out with the second press. In the stop mode, the tray slides out with the first press. Press the key while the disc tray is open to close the tray.

When the *Play key* is pressed after placing a disc on the disc tray, the disc slides into the player and the play mode begins. If the player is in the stop mode with a disc inserted in the player, pressing the Play key activates the play mode. Pressing Play during play restarts play from the beginning of the current chapter or track. Pressing Play can also start a programmed or FTS play sequence. If the player is in a mode other than normal play (strobe, freeze, or still), pressing the Play key returns the player to normal play.

While playing a CDV or the video portion of a CDV Single, pressing the *Pause key* freezes the picture. To release the freeze mode, press Pause again or press the Play key. While playing a CD or the audio portion of a CDV Single, pressing the Pause key stops play temporarily. To release the pause mode, press Pause again or press the Play key.

Press one of the *Skip keys* to skip to the beginning of a chapter or track while in the play mode. The Skip keys also function as program-check keys, which allow the user to verify the content of the program being entered. When clearing an FTS program, the Skip keys are used to recall the ranking numbers.

When one of the *Scan keys* is pressed and held during play, play is scanned forward or backward. The scanning speed varies in two steps (low speed for the first 2 s, then high).

Pressing one of the *Index-Skip keys* activates an index-skip operation with a

CD or CDV on which the index numbers have been recorded. Pressing an Index-Skip key repeatedly skips the same number of indexes as the number of times the key is pressed.

The *Multiple-Speed Play keys* are used only for CDV or the video portion of CDV Single. The audio content is muted when the multiple speed is used, and the user may vary the video-play speed as desired (in the range from 3 s per frame to 3 times the normal play speed).

When power is switched on, the initial speed selection is set to one-quarter of the normal play speed. The speed can be switched in nine steps by pressing the *Multiple-Speed Set keys* (labeled + and −). The speed is displayed on the TV screen (OSD) as long as the Multiple-Speed Set keys are pressed.

Press one of the *Still/Step keys* (forward or reverse) to freeze (still frame) the picture. The still frame can be either advanced to the next frame by pressing the Forward-Step key, or the previous frame can be displayed by pressing the Reverse-Step key. Press the Play key to cancel the still picture.

The *D-A/CX key* is available only for standard CDV discs (8 and 12 in) and is used to select the audio signal that is sent to the rear-panel Digital/Analog audio jacks. The front-panel *digital-sound* display is on when the disc being played contains a digital audio signal.

When the D-A/CX key is pressed, the digital-sound display goes out, indicating that analog audio is selected as the output signal (at the Digital/Analog jacks). The D-A/CX key can also be used to toggle the CX noise-reduction system on and off.

The *Bilingual key* is used when a bilingual disc is played, to select among left (1/L), right (2/R), or both channels.

The *OSD key* is used to toggle the on-screen display on and off.

The *Select key* is active only when playing a standard CDV. Press the Select key to recall the specific position you want to view and/or listen to (search operation). With a CAV (standard play) disc, pressing the select key activates the *frame-number search mode*. When playing a CLV (extended play) disc, pressing the Select key activates *time-number search mode*. The *Chapter-Select* mode may be activated for either CAV or CLV with the Select key.

The *Program key* is used to store chapters or tracks in a different order than on the disc. After the sequence is programmed, the Play key is pressed to activate the programmed play sequence.

The *Store key* is used to enter the chapters or tracks for a programmed play. It is also used to enter an FTS program.

The *Recall key* is used to verify or modify the programming content. It is also used to verify the FTS program content.

The *Numeric keys* are used when searching or programming the chapters or tracks.

The *Clear key* is used to cancel repeat play, interrupt a search operation, correct an entry made using the numeric keys (holding the key down clears the digit entry operation), cancel programmed play, cancel random play, and clear an FTS program.

The *M1:FTS key* is used to program or play an FTS sequence for either CD or CDV discs.

The *M2:Random-Play key* is used to start random play (active only with CD and the audio portion of CDV Single).

The *M3:TV/CDVP key* is used to switch between the disc-play and TV-broadcast signals.

The *M4:A-B key* is used to repeat a block of signals between points A and B.

The *M5:Repeat key* is used to repeat the play of a track or chapter or to repeat the entire side of the disc being played.

The *Picture Memory* key is active with CDV discs only. Press Picture Memory to store the current picture (frame) in memory. After playback, the stored picture is displayed on the TV screen while you are exchanging discs. Also, the stored frame is displayed while the audio portion of a CDV Single is playing.

The *Picture Effect* key is active with CDV discs only, and it allows the user to display a special-effects picture by dividing the picture into many small fragments or a mosaic.

The *Freeze key* is active with CDV discs only, and it stops the picture (freeze mode) while the sound continues.

The *Strobe key* is active with CDV discs only. When the Strobe key is pressed during playback, the "strobe motion" play is activated. After pressing the Strobe key, adjust the required interval between pictures with the Multiple-Speed Set keys.

The *Display* key is used when the display indications on the player and/or the on-screen indications on the TV screen are not required during playback.

The *Jog-On/Off* key is used to activate the *jog dial*. Press Jog-On/Off once to activate the jog-dial capability, which is shown when the *jog-mode* indicator turns on. Press Jog-On/Off again to release the jog mode. The playback speed can be continuously varied with the *jog dial/shuttle ring*.

When the remote unit is not set to the CDV mode, press the CDV key to recall the CDV mode. When CDV is pressed, CDV appears in the remote-control display.

The *contrast control* is used to adjust the contrast of the remote-unit display.

The *Balance* keys are provided for operating audio components or TV sets with an audio-balance function.

CHAPTER 4
TEST EQUIPMENT, TOOLS, AND ROUTINE MAINTENANCE

This chapter describes the test equipment and tools you will need for CD and CDV player service. We also discuss routine maintenance for CD and CDV players. Remember that the information in this chapter is general in nature. If you are going to service a particular player, get all the service information you can on that player. Likewise, if you plan to go into CD and CDV player service on a grand scale, study all the applicable service instructions you can find; then, when all else fails, you can follow instructions. We discuss adjustments (mechanical) using tools and test equipment in Chap. 7. We also discuss the use of tools and test equipment for adjustment and troubleshooting in Chaps. 8 and 9.

4.1 SAFETY PRECAUTIONS DURING SERVICE

In addition to a routine operating procedure (for both test equipment and the player), certain precautions must be observed during operation of any electronic test equipment during service. Many of these precautions are the same for all types of test equipment; others are unique to special test instruments, such as meters, scopes, and signal generators. Some of the precautions are to prevent damage to the test equipment or to the circuit on which the service operation is being performed. Other precautions are to prevent damage to you.

4.1.1 General Service Safety Precautions

It is assumed that you are already familiar with general safety precautions regarding electronic service. Such precautions include (but are not limited to) checking the warning symbols on test equipment (a *triangle with an exclamation point* at the center for controls that require special precautions and a *zigzag line simulating a lightning bolt* for high voltages), checking metal cases for leakage and grounds, handling high voltages with care (fortunately the line voltage is usually the highest in CD and CDV players), removing power before making connections, using an *isolation transformer*, avoiding vibration and mechanical shock (particularly to the optical components), studying the circuits for connecting test equipment, and making *leakage current tests*.

4.1.2 Basic Handling and Service Precautions

The following basic safety precautions should be studied thoroughly and then compared with any specific precautions called for in the test-equipment or player service literature and in the related chapters of this book.

Moisture Condensation. Try to avoid servicing the player immediately after moving it from a cold to a warm place or soon after heating a room that was cold. Either of these conditions can cause moisture condensation. Excessive condensation (which is rare) can cause possible damage to circuits. More likely, *condensation can fog the lenses in the optical system*. You can clean the surface of the objective lens (with a clean, soft, dry cloth), but the remaining lenses are not accessible. You must wait until the condensation evaporates from the internal lenses. You can also try turning on the power (but not operating the player) and allowing heat from the transformers to remove the moisture.

Handling and Storage. Avoid servicing players in the following places: extremely hot, cold, humid, or dusty areas; near appliances that generate strong magnetic fields (or ones that are affected by such fields); places subject to vibration; and poorly ventilated places. Do not block the ventilation openings on the player. Do not place anything heavy or anything that might spill on the player. Use an accessory cover (if available) to prevent dust and dirt from accumulating on the player before or after service. Use the player in the horizontal (flat) position only. Do not lubricate player motors or any point not recommended for lubrication in service literature. *Generally, very little lubrication is required for CD and CDV players*.

When reassembling any player, always be certain that all the protective devices (shield plates, etc.) are put back in place. When servicing or testing is required, observe the original wire routing. *Pay particular attention to the wiring associated with the disc compartment, doors, and optics*. Since these components are subject to *constant movement, wire routing is critical*. For example, wires can be caught in gears or subjected to excessive strain if not properly routed.

Always follow the packing and shipping instructions (if any) found in the service literature. Always use the transit or shipping screw described in Chap. 3 to hold the optics in place during transit.

Interlocks. Do not defeat any type of interlock on a player (at least not permanently). If you must override an interlock during service (try to avoid this), *do not permit the player to be operated by others without all protective devices correctly installed and functioning*. Servicers who defeat safety devices or fail to perform safety checks may be liable for any resulting damage.

Special Product Safety Notices. Many electrical and mechanical parts in CD and CDV players have *special* safety-related characteristics, some of which are often not evident from visual inspection. Likewise, the protection provided by such parts cannot necessarily be obtained by replacing the parts with components rated for higher voltage, wattage, and so on. The manufacturers identify such parts in their service literature.

One common means of identification is *shading on the schematics and/or parts lists*, although not all manufacturers use shading, nor do they limit identification to shading. For example, some manufacturers use a *dark black pattern* on those areas of the printed circuit (PC) pattern that require special care in repair. Other manufacturers use an *exclamation point within a triangle* for critical parts.

Always be on the alert for any special product safety notices, special parts identification, and the like. Use of a substitute part that does not have the *same safety characteristics* (not just the same electrical or mechanical characteristics) might create shock, fire, and/or other hazards. A simple way to solve the problem is to use the part recommended in the service literature.

Altering the Players. Design alterations, including (but not limited to) addition of auxiliary audio output connections, cables, accessories, and the like, might alter the safety characteristics of the player and create a hazard to the user. Do not alter or add to the mechanical and/or electrical design of players. Any design alterations or additions may void the manufacturer's warranty and make the servicer responsible for personal injury or property damage.

Professional Electronic Service Practices. The author assumes that you are already familiar with good electronic service practices (removing or disconnecting the power cord before replacing PC boards and parts, installing heat sinks as required on solid-state devices, connecting test-instrument ground leads to player grounds before connecting the test instrument, and so on). The author also assumes that you can handle electrostatically sensitive (ES) devices [such as field-effect transistors (FETs), metal-oxide semiconductor (MOS) and complementary MOS (CMOS) chips, etc.]; that you can solder and unsolder integrated circuits (ICs), transistors, diodes, and the like, including leadless components and surface-mount design (SMD) devices; and that you can repair PC boards as needed. If any of these seem unfamiliar to you, please, please do not attempt to service any CD or CDV player.

4.1.3 Laser Safety

CD and CDV players have a laser diode (or laser tube, in the case of older CDV players) which creates two possible service hazards. First, both the laser diode and tube produce a *potentially dangerous light beam*. As in the case of any other very intense light source, direct exposure to a laser beam can *cause permanent eye injury or skin burns*. Second, the light beam produced by a solid-state laser diode is *invisible*. (This is in contrast to the red light beam produced by a laser tube.) Since the diode beam is invisible, you are never quite sure when the beam is present.

CD and CDV players are designed to be operated without the operator being exposed to the beam. This is also essentially true for the servicer, with one major exception. If you gain access to the laser (by removing covers, opening the disc compartment, etc.) and keep power on the laser (by overriding interlocks, etc.), the beam may get you. None of this should frighten you, but the problems should keep you on your toes when servicing. It is the servicer's job to exercise all caution to avoid any direct exposure to the laser beam.

U.S. federal law (and the laws of most countries) requires that servicers be advised of possible laser dangers. There is at least one warning label on all players, and often more than one (typically one on the outer cover and one in the disc compartment near the objective lens). The label reads something like "CLASS 1 LASER PRODUCT: Product complies with DHHS rules CRF subchapter J part

1040;10 at date of manufacture; DANGER: Invisible laser radiation when open and interlock failed or defeated. AVOID DIRECT EXPOSURE TO BEAM." On CD players designed for Canadian use, the warning label is strengthened by a *light-burst pattern within a triangle*.

Always be on the alert for these warning labels when servicing a CD or CDV player. Equally important, make sure all shields and covers are in place and that interlocks are working *before* you turn the player over to the customer.

In addition to producing a potentially dangerous beam, *lasers produce strong electromagnetic radiation*. This is not usually harmful to people but can be disastrous to magnetic tape, some wristwatches, and anything else affected by magnetic fields (possibly pacemakers). Do not have magnetic tape, audio or video cassettes, or any other magnetic device nearby when servicing a CD or CDV player, particularly when the player covers and shields are off. To be on the safe side, keep all magnetic tape away from CD and CDV players, even with all covers in place.

4.1.4 Laser Checks and Adjustment

Most CD and CDV player manufacturers recommend some means of checking the laser without having to monitor the beam with a light meter (as was recommended on some older CDV players). Study the service literature for the CD player you are servicing to find the recommended procedure. In the absence of any specific recommendations, here are some tips on checking lasers.

Even though the laser beam is invisible, the diffused laser beam is usually visible at the objective lens (the lens appears to glow when the beam is on). Also, when power is first applied to the optical circuits of most CD players, the objective lens moves up and down (usually three times, as discussed in Chap. 5) to focus the beam on the disc. So, if you apply power and see the objective lens moving, it is reasonable to assume that the laser is on and producing enough power.

The check procedure brings up some obvious problems. First, on most players, if you open the disc compartment and gain access to the objective lens, you must override at least one interlock. Next, many players have some provision for shutting down the player optics if there is no disc in place, so you must override this feature.

Most important, *never, never look directly into the objective lens with the power applied*. Keep your eye at least 30 cm (12 in) from the lens. The purpose of the objective lens is to focus the beam sharply onto the disc. The lens can also focus the beam sharply into your eye.

Figure 4.1 shows the recommended procedure for both the laser-diode check and focus-search check on a typical CD player. For both checks, you must remove the outer cover and locate the objective lens.

To check the laser diode, ground the laser on/off control signal from the system-control microprocessor (such as pin 51 of IC_{301}, shown in Fig. 5.3), and check for diffused light at the objective lens.

To check the focus search (Fig. 5.6) block the disc detection phototransistor with paper (to simulate a disc in place) and press the Play button. Check that the laser emits light and that the objective lens moves up and down (in response to focus up-down (FUD) signals from pin 50 of IC_{301}, shown in Fig. 5.6). Note that the objective lens may not always move the same number of times or by the same amount. (In some players, you can barely see the lens movements.)

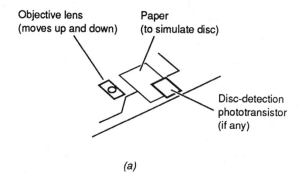

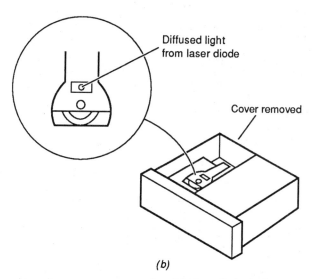

FIGURE 4.1 (*a*) Focus-search and (*b*) laser-diode checks.

Some manufacturers recommend that you *monitor the eight-to-fourteen modulation (EFM) signal* (or some similar signal) to check the laser diode. If the EMF signal is normal or can be adjusted to a normal value, the laser diode must be functioning normally. We describe such checks and adjustments in Chaps. 8 and 9.

4.1.5 Replacement of Optical System Components

As with many electronic components, the laser diode can suffer *electrostatic breakdown* because of the potential difference generated by the charged electrostatic load on clothing and the human body. The laser diode is usually considered part of the optical system or pickup assembly, and most player manufacturers recommend replacement of the complete pickup assembly as a package. Most manufacturers do not supply individual parts for the optical system.

Figure 4.2 shows recommended procedures for handling the optical system of a typical CD player when the entire assembly is to be replaced. The following notes supplement the procedures shown.

Place a conductive sheet on the workbench. In some players, the *black sheet used as the repair parts wrapping* is a conductive sheet.

Place the player on the conductive sheet so that the chassis touches the sheet. This makes the chassis (and pickup assembly) the same potential as the conductive sheet.

Place your hands on the conductive sheet. This makes your hands the same potential as the sheet.

Remove the optical-system block from the bag (which is made of conductive material).

Perform the necessary work on the conductive sheet. Be careful that clothing does not touch the optical-pickup block.

If practical, use a wrist strap (Fig. 4.2) with an impedance-to-ground of less than 1 MΩ. The workbench and/or conductive sheet should also have an impedance-to-ground of less than 1 MΩ. Remember that static electricity builds up on clothing and is often not fully drained off, even with wrist straps and grounded workbenches.

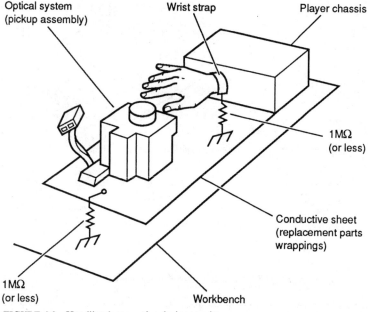

FIGURE 4.2 Handling laser optics during service.

4.2 TEST EQUIPMENT

The test equipment used in CD and CDV player service is basically the same as that used in TV, VCR, and audio service. That is, most service procedures are performed using meters, generators, scopes, distortion meters, power supplies, and assorted clips, patch cords, and so on. Theoretically, all CD and CDV player service procedures can be performed using conventional test equipment, provided that the generators cover the appropriate frequencies, the scopes have the necessary gain, and so on.

For these reasons, we do not go into test equipment in great detail. If you have a good set of test equipment suitable for conventional TV, VCR, and audio work, you can probably service any CD or CDV player. However, there are certain items that can be most useful in CD and CDV service: *test discs, shop-standard stereo amplifier with speakers*, and a *receiver/monitor*. A *distortion meter* is also helpful but not absolutely essential.

4.2.1 Test Discs

Many CD and CDV player manufacturers provide test discs (also known as *check discs, reference discs*, or *alignment discs*) as part of their recommended test equipment and/or tools. Some manufacturers even recommend the test disc of another manufacturer.

A test disc is essentially a standard disc with several very useful signals recorded at the factory using very precise test equipment and signal sources. You play the test disc on a player being serviced and note the response and/or use the signals to perform alignment and adjustment of the player. With the proper test disc, you can often eliminate your own signal sources (signal generators, audio generators, etc.).

One major problem with test discs is the lack of standardization. You will probably need several test discs if you service many different models of CD and CDV players. The alignment procedures found in most service literature call for signals not available on all test discs. The only way around this problem is to *use the recommended test disc in all cases*. Of course, you can use any known-good disc for a final, after-service check of the player, but this does not give you the necessary signals to perform the adjustment procedures.

4.2.2 Receiver/Monitor

If you are planning to go into CDV service full time, you should consider a receiver/monitor such as is used on studio and broadcast video work. These receiver/monitors are essentially TV receivers with video and audio inputs brought out on some accessible point (usually the front panel). There are also many TV sets with monitor inputs (but not outputs) currently on the market.

The output connections of an industrial or studio monitor make it possible to monitor broadcast video and audio signals as the signals appear at the output of TV receiver intermediate frequency (IF) section (the so-called *baseband signals*, typically in the range 0 to 4.5 MHz, at 1 V for video and 0 dB or 0.775 V for audio). These output signals from the studio monitor are used in the service of other video equipment (such as VCRs) and are not generally useful for CDV service.

The *input connections* of an industrial monitor or a monitor-type TV are very useful in CDV service. The input connections make it possible to inject video and audio signals from the player [before the signals are applied to the radio frequency (RF) output unit of the player] and monitor the display. Thus, the baseband output of the player can be checked independently from the RF unit.

If you do not want to use either a studio monitor or monitor-type TV, you can use a standard TV receiver to monitor the player. Of course, with a TV, the player audio and video signals are used to modulate the player RF unit. The output of the RF unit is then fed to the TV antenna input. Under these conditions, it is difficult to tell if faults are present in the player audio or video or in the RF unit. Similarly, if you use a signal generator for a video source, the generator output is at an RF or IF frequency, not at the baseband frequencies.

If you use a TV as a monitor, it is often helpful to adjust the vertical height control of the TV to *underscan the picture*. In this way, you can easily see the video switching point in relation to the start of vertical blanking (an old VCR service trick).

4.2.3 Shop-Standard Stereo Amplifier with Speakers

If you are already in audio and stereo service work, you probably have at least one shop-standard amplifier and speaker system. Any system suitable for final checkout of turntables, tape players, and so on will do the job for a CD or CDV player. Remember that the CD player output to the amplifier is about 2 V (either variable or fixed, depending on the player).

4.2.4 Distortion Meters

If you are in audio and stereo service, you probably have distortion meters (and know how to use them effectively). This is not true for many TV and VCR service shops. As a practical matter, you can probably get by without a distortion meter in CD and CDV player service. If distortion is severe, you will hear it. If the distortion is below a level where it can be heard, you can generally forget the problem. Of course, if you have a customer with a "golden ear," an accurate distortion meter is an excellent tool for settling "discussions" concerning the player's performance, especially after service. (Note that the owners of expensive stereo equipment all have golden ears.)

4.3 TOOLS

Most CD and CDV players require some form of special tools for full service. However, there is no standardization in this area. For example, the service literature for the player of one manufacturer recommends such items as a disc chuck, platter eccentricity gauges, gain and position adjustment jigs, eccentric screwdrivers, and thickness gauges, in addition to four test discs. Another manufacturer of an equally complex player recommends only one height-adjustment

TEST EQUIPMENT, TOOLS, ROUTINE MAINTENANCE   4.9

fixture, plus one special test disc. Still another manufacturer recommends no special tools of any kind for mechanical adjustment or for disassembly and reassembly but calls for one special test disc. For these reasons, we do not go into special tools in this section. The tools required for test and adjustment are discussed in the related chapters (7 through 9). However, there are some things to remember if you plan to go into CD and CDV service on a full-time basis.

Study the service literature and use the recommended tools. If the tools appear in the service literature, they should be available from the manufacturer (or the service literature will show how to fabricate the tool).

In most cases, there are other tools and fixtures used by the manufacturers for both assembly and service of players at the factory. These factory tools are not available for field service (not even to factory service centers, in some cases). This is the manufacturer's subtle way of telling service technicians that they should not attempt any adjustment (electrical or mechanical) not recommended in the service literature.

The author strongly recommends that you take this hint. There are many horror stories told by factory service people concerning "disaster area" players brought in from the field. Most of these problems are the result of tinkering with mechanical adjustments (although there are some technicians who can destroy a player with a simple electrical adjustment). One effective way to avoid this problem is to *use only recommended factory tools and perform only recommended adjustment procedures*.

In addition to possible special tools, the mechanical sections of CD and CDV players are disassembled, adjusted, and reassembled with common hand tools, such as wrenches and screwdrivers. Most players are manufactured to Japanese *metric standards*, and your tools must match. For example, you will need metric-sized Allen wrenches and Phillips screwdrivers with the Japanese metric points.

4.4 PERIODIC MAINTENANCE AND DISC CARE

There is considerable disagreement among CD and CDV player manufacturers concerning the need for periodic maintenance or routine checks. At one extreme, a certain manufacturer recommends replacement of a few parts after a given number of playing hours or playing times. (They recommend that the optical pickup be replaced at 5000 h of playing time and all motors be replaced after 10,000 plays.) At the other extreme, another manufacturer recommends no periodic replacement, cleaning, lubrication, or adjustment of any kind. "Fix it if it breaks down" is the rule. (It is fair to say that this rule will probably be observed religiously.)

Somewhere between these two extremes, other manufacturers recommend adjustment (electrical and/or mechanical) *only as needed* to put the player back in service or when certain parts or assemblies have been replaced. However, most player manufacturers recommend a complete checkout, using the appropriate test discs, after service.

The author has no recommendations in the area of periodic maintenance except that you follow the manufacturer's recommendations. The author also realizes that the general public regards a CD or CDV player in the same way it does an audio or stereo component or TV (that is, "bring it in for service when it stops working"). However, here are some maintenance procedures that you can pass along to your customers.

Cleaning the Player. Use a soft, clean cloth to wipe off dust and dirt accumulated on the player. If absolutely necessary, moisten a soft cloth with diluted neutral detergent to remove heavy dirt. *Never use* paint thinner, benzene, or other solvents since any of these solvents can react with the player surfaces and cause color changes (or even melting).

Objective Lens Care. The objective lens is a key part of a CD or CDV player. The lens surface must be clean (and free of moisture) to maintain the best performance. *Try not* to touch the lens surface. Keep the disc compartment closed to prevent dust and dirt from accumulating on the lens. Open the disc compartment only when inserting and removing the disc. If excessive dust or dirt accumulates on the objective lens, sound and/or picture quality can be degraded. Dust can be removed from the objective lens with an air blower (such as used for camera lenses).

Basic Disc Care. Since both CD and CDV discs are played without physical contact to the playing surface, the discs do not wear out or degrade in quality. Also, minor scratches or small amounts of dust or fingerprints have little effect on sound and/or picture. A heavy accumulation of dust and dirt, however, can reduce sound and picture quality and should be removed by wiping both sides of the disc with a clean, soft, dry cloth. Try to *handle all discs by their edges*, the same as a phonograph record or LP.

Disc Cleaners. The author has no recommendations for or against commercial disc cleaners. A soft, dry cloth will remove most dirt found on discs. However, if a disc is exposed to very dirty fingers, liquids, tobacco smoke, fire and brimstone, and so on, you may need help in restoring the disc. There are even cases where a mold has started to form or the disc surface has oxidized. Cleaners can be of great help in these extreme cases.

Removing Scratches from a Disc. Use the following procedure to remove scratches from the surface of a CD or CDV. *Use a polishing compound* such as Simonize Pre-Softened White Polishing Compound. *Do not use a rubbing compound*. Place a small amount of polishing compound on a dampened, soft cloth. With a *light*, circular motion, rub the surface of the disc until the scratched or marred area is restored to a luster. After the marred area is restored, remove any residue from the disc surface with a clean dry cloth. If this procedure does not remove the scratch or mar, the disc is probably beyond hope.

Cold Discs. When discs are subjected to extremely cold temperatures (try to avoid this), allow about an hour for the disc to return to room temperature before playing.

Warped Discs. There are a variety of playback problems associated with warped CD and CDV discs. Typical problems include hanging (or sticking), skipping, and a loss of focus. A severely warped disc can even hit the player lid or objective lens. To determine if a disc is warped, lay the disc on a flat surface and press down on the center. If no movement is detected, turn the disc over and again press down in the center. If movement is detected on either side, measure the amount of movement. For a 12-in disc, the movement should not exceed about 3 mm (just under $1/8$-in). If the amount of warpage is over 3 mm, the disc is defective and should be replaced.

CHAPTER 5
TYPICAL CD PLAYER AND CD-ROM CIRCUITS

This chapter describes the theory of operation for a number of CD player circuits. The circuits described here include all circuits associated with mechanical components, laser optics, laser signal processing, audio, autofocus, laser tracking, CD turntable motor, and antishock. The chapter concludes with a discussion of CD-ROM circuits, concentrating mostly on the differences between CD and CD-ROM circuits.

After studying the circuits found in this chapter, you should have no difficulty in understanding the schematic and block diagrams of similar CD players. This understanding is essential for logical troubleshooting and repair, no matter what type of electronic equipment is involved. No attempt has been made to duplicate the full schematics for all circuits. Such schematics are found in the service literature for the particular CD player.

Instead of a full schematic, the circuit descriptions are supplemented with partial schematics and block diagrams that show important areas such as signal-flow paths, input-output, adjustment controls, test points, and power-source connections (the areas most important in service and troubleshooting). By reducing the schematics to these areas, the circuits are easier to understand, and you will be able to relate circuit operation to the corresponding circuit of the player you are servicing.

5.1 RELATIONSHIP OF CD PLAYER CIRCUITS

Figure 5.1 shows the relationship of player circuits. The laser beam is generated by laser diodes located on the pickup assembly. The beam is passed through a lens assembly and focused on the CD. The reflected beam from the CD is then applied to phototransistor sensors through the lens assembly.

The low-amplitude output of the sensors is applied through preamps, filters, and waveshaping circuits IC_{601} through IC_{603} back to the laser drive IC_{604}, as described in Sec. 5.3. The output is also applied to data-strobe, signal-generator, and error-correction circuits IC_{401} through IC_{403}.

The purpose of the data-strobe circuit (Sec. 5.4) is to correctly identify whether the signals have a digital status of 0 or 1 and to clock the sync, control, and audio information. Once this is done, the circuits perform signal processing (error correction and decoding) to produce four major outputs.

Two of the outputs are left and right audio signals, as described in Sec. 5.5. A

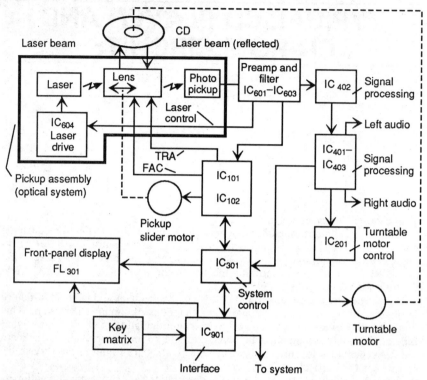

FIGURE 5.1 Relationship of CD player circuits.

third output is the necessary control information for the CD motor (turntable) circuits (Sec. 5.8). The fourth output represents information about where the laser beam is located on the CD and is applied to the system-control microprocessor IC_{301}. In turn, IC_{301} generates signals applied to servo control IC_{101} and IC_{102}.

The servo circuits IC_{101} and IC_{102} are responsible for moving the laser pickup assembly across the CD, keeping the beam in focus, and keeping the focus properly centered on the CD tracks, as described in Secs. 5.6 and 5.7.

IC_{301} and IC_{901} provide for interfacing among the player circuits. IC_{901} is responsible for communications between the CD player and the audio system (if any). IC_{901} also receives and decodes all inputs from the front-panel keyboard matrix, transferring such commands to both FL_{301} (for front-panel display) and IC_{301} (for system control).

5.2 MECHANICAL FUNCTIONS

Figure 5.2a shows the major mechanical components of our CD player. Figure 5.2b shows the associated circuits. The mechanical components of a CD player perform two major functions: loading and unloading the CD and driving the op-

TYPICAL CD PLAYER AND CD-ROM CIRCUITS 5.3

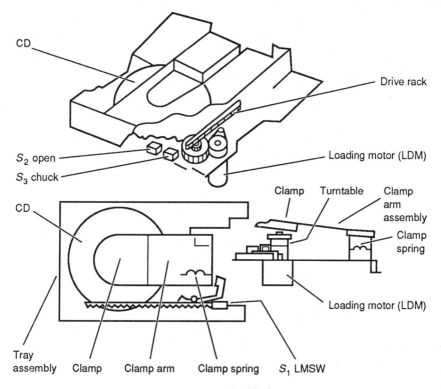

FIGURE 5.2 (a) Major mechanical components of a CD player.

tical pickup (and laser beam) across the CD. Chapter 7 includes much more detailed descriptions of the mechanical functions and components.

5.2.1 Loading and Unloading

In our player, the tray is opened by a loading motor, or LDM (in response to pushing the Open/Close button), a CD is inserted within the tray, the tray and CD are placed back in the player (by the LDM), and then the CD is installed on the turntable (by the same drive mechanism and LDM).

Operation of the LDM is controlled by limit switches S_2 and S_3 and by IC_{301}. So when you are troubleshooting any mechanical functions, you must study both the mechanical drawing and circuits. The following paragraphs describe both the mechanical and electrical functions of our player. Again, remember that these descriptions must be compared with the descriptions found in (or omitted from) service literature.

Note that most of the mechanical components are part of a mechanism secured to the left side of the player mainframe by two rails. The tray is moved out of the player front panel on rollers by the LDM. This action also raises the spring-loaded clamp. A CD is installed manually on the tray, and the tray is

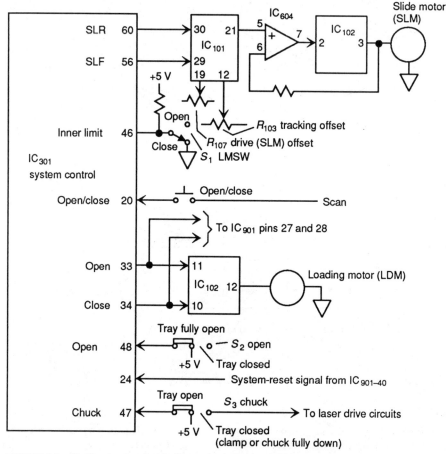

FIGURE 5.2 (b) Circuits associated with mechanical components.

pulled within the player by the LDM. This action also lowers the clamp so that the CD is pressed against the turntable motor.

The LDM receives open and close drive signals from IC_{301} through IC_{102}. In turn, IC_{301} receives indicator signals from the tray-open switch (open) S_2 and the tray-closed (chuck) switch S_3. (In the service literature, the tray-open switch S_2 is called the *open* switch for obvious reasons. The tray-closed switch S_3 is called the *chuck* switch since the switch is actuated only when the CD is clamped or "chucked" onto the turntable.)

Note that the open and close drive signals from IC_{301} are also applied to IC_{901}, at pins 27 and 28. This informs IC_{901}, and the system (if any), that the tray is moving in or out or has stopped.

When the tray is open and the front-panel Open/Close button is pushed, IC_{301} produces a close signal at pin 34. This signal is applied to the LDM through IC_{102} and causes the LDM to close the tray. Until the tray is in and the clamp is fully down, S_3 remains in the tray-open position. This applies +5 V to the laser drive circuits, as described in Sec. 5.3.

When the tray is closed and the front-panel Open/Close button is pressed, IC_{301} produces an open signal at pin 33. This signal is applied to the LDM through IC_{102} and causes the LDM to open the tray.

Until the tray is fully out, S_2 remains in the tray-closed position. This leaves pin 48 of IC_{301} open, telling IC_{301} to continue applying an open signal. When the tray reaches the fully open position, S_2 actuates (moves to the tray fully open position). This applies +5 V to pin 48 of IC_{301}, and the LDM stops driving the tray.

5.2.2 Optical Pickup (Beam) Drive

A motor is required to keep the beam moving across the CD at a constant rate, even though the CD speed changes (as described in Sec. 5.8). In our player, the optical pickup (also called the *slide* or *sled* in some literature) is moved across the CD by the slide motor (SLM). In turn, the SLM is operated by slide-forward (SLF) and slide-reverse (SLR) signals from pins 56 and 60 of IC_{301} through IC_{101}, IC_{604}, and IC_{102}.

Note that R_{107} provides an offset adjustment for the drive signals applied to the SLM. The adjustment sets the point where the pickup accesses the beginning of the CD (the directory). If the adjustment is not correct, the program information may not be read properly. R_{103} provides an offset adjustment for the tracking signals applied to the SLM and is adjusted to produce the best response (minimum audio dropout), as described in Chap. 8.

When a CD is first installed (with the tray fully in and clamp fully down), IC_{301} produces a temporary SLR signal at pin 60. (The temporary SLR is generated by a system-reset signal from pin 40 of IC_{901}.) The SLR is applied to the SLM and causes the SLM to move to the limit of the CD (the CD start position). As you probably know, CDs start playing from the center and play to the outside, opposite to the play of an LP record.

Until the pickup and beam are at the inner limit, S_1 remains in the open position. This applies +5 V to pin 46 of IC_{301}, telling IC_{301} to continue applying the temporary SLR signal to the SLM. When the pickup reaches the inner limit, S_1 actuates (moves to the closed position). This grounds the signal at pin 46 of IC_{301}, the SLR signal is removed, and the SLM stops.

When the pickup is resting at the inner-limit position and the front-panel play button is pressed, IC_{301} produces an SLF signal at pin 56. The SLF is applied to the SLM through IC_{101}, IC_{604}, and IC_{102} and causes the SLM to drive the pickup and beam across the CD.

5.3 LASER OPTICS AND CIRCUITS

Figure 5.3 shows the laser optics and circuits. These components are responsible for recovering, performing preliminary preamplification, and waveshaping the audio signal recovered from the CD. Before we get into the circuits, let us review the basic functions of laser optics (as applied to CD players) that were discussed in Chap. 1.

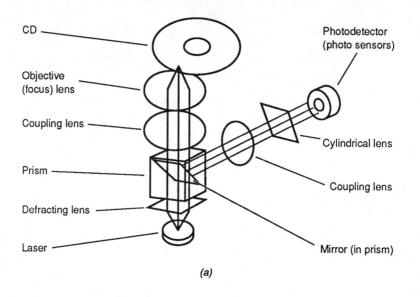

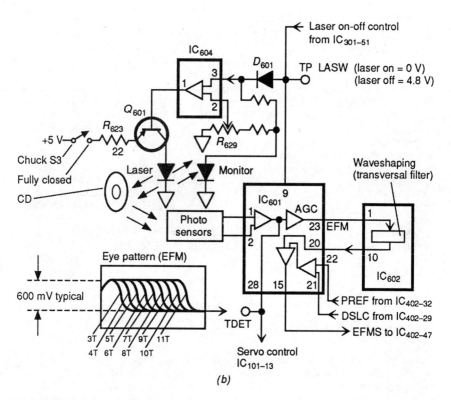

FIGURE 5.3 Laser optics and circuits.

5.3.1 Laser Optics

As shown in Fig. 5.3a, the solid-state laser used in CD players emits a beam that is first passed through a defracting lens. This lens splits the main laser beam into three separate beams, which are next passed through a prism, a coupling lens, and an objective (or focus) lens to the CD. The light reflected back from the CD is first passed through the objective and coupling lenses to the prism. The beams are then reflected (by a mirror in the prism) to a photodetector through another pair of coupling and cylinderical lenses. The purpose of the cylindrical lens is to cause the normally round laser beam to distort in an oblong fashion. The beam is elongated if the objective lens is not properly focusing the beam on the underside of the CD (as discussed in Sec. 5.6).

The photodetector serves three primary purposes: (1) to generate the signals necessary for autofocus, (2) to generate the tracking signals that allow the beams to follow the tracks on the CD accurately, and (3) to generate the digital audio signal that is converted back to an analog audio signal.

5.3.2 Laser Control

The conduction rate of the laser diode is controlled by driver Q_{601}. Note that Q_{601} receives emitter current through R_{623} and the chuck switch S_3. As discussed in Sec. 5.2, S_3 is actuated only when the tray is fully in and the clamp is fully down. This removes 5 V from pin 47 of IC_{301} and applies 5 V to the emitter of Q_{601}, turning Q_{601} on.

With this safety circuit, the laser remains off until the CD tray is fully closed and the CD is in place to receive the laser beam. Once power is available to the laser through S_3 and Q_{601}, the laser is turned on and off by a control signal from pin 51 of IC_{301}.

When pin 51 of IC_{301} goes high (laser off), the high is applied to pin 3 of IC_{604} through D_{601}, causing pin 1 of IC_{604} to go high and turning off Q_{601}. This removes power from the laser diode. With pin 51 of IC_{301} low (laser on), the low is applied to pin 3 of IC_{604} through D_{601}, causing pin 1 of IC_{604} to go low and turning Q_{601} on. This applies power to the laser diode.

As discussed in Sec. 5.3.1, the laser beam is reflected from the CD onto the photosensor. The beam is also applied directly to a monitor diode that regulates the amount of current through the laser diode (and thus the intensity of the laser beam). Most CD players include circuits to monitor and control the amount of light emitted by the laser. This is necessary for proper performance of the optical system. For example, a low output from the laser diode can produce tracking errors as well as audio dropout.

The output of the monitor diode is applied to the input at pin 3 of IC_{604}. The other (inverting) input at pin 2 receives an adjustable reference voltage, set by R_{629}. The output of IC_{604} is applied to the laser diode through Q_{601}. If the laser diode output goes below the desired reference level, the monitor diode output decreases and the IC_{604} output goes less positive. This increases the laser drive current supplied by IC_{601}, increasing the laser diode output back to normal. The opposite occurs if the laser output increases. The laser diode output can also be set to an optimum value with R_{629}.

5.3.3 Laser Adjustment

Always follow the service-literature instructions when adjusting the laser diode. Typical adjustment procedures are described in Chap. 8.

5.3.4 Laser Beam Processing

The laser output is reflected from the CD onto the photosensors. The outputs of the photosensors are applied to pins 1 and 2 of IC_{601}. Within IC_{601}, the outputs are applied to a differential amplifier. The output from the differential amplifier is applied to pin 13 of servo control IC_{101}, through pin 28 of IC_{601}, and to an amplifier within IC_{601}.

The signal at pin 28 of IC_{601} is known as the main eight-to-fourteen modulation (EFM) signal and represents the primary signal output from the CD optics. As discussed in Chap. 1, this signal is also called the *high-frequency* (HF) signal or the *RF* signal and produces what is known as the *eye pattern* shown in Fig. 5.3b.

Although the EFM signal appears to be a series of sine waves at this point (at the TDET test point), the signals are digital. The designations 3T, 4T, and so on refer to 3 times the period required to read a pit (Sec. 5.7) on the CD, 4 times the period, and so on. The limits of 3T to 11T are set by CD specifications.

The EFM signal also exits IC_{601} (at pin 23) through an automatic gain control (AGC) amplifier and enters waveshaping (transversal filter) circuits within IC_{602} through pin 1. The waveshaping circuits assure that the 3T signal, or high frequency, is equal in amplitude to the 11T signal, or low frequency. The waveshape output is applied to the signal-processing circuit at pin 47 of IC_{402} through pin 10 of IC_{602}, pin 20 of IC_{601}, a comparator within IC_{601}, and pin 15 of IC_{601}. The main signal is then processed in IC_{402}, as described in Sec. 5.4.

The comparator within IC_{601} shapes the EFM signal into square waves (known as the EFM square wave signal, or EFMS). The EFM signal is compared to a dc threshold voltage developed by circuits within IC_{402} and IC_{601}. In brief, the EFMS is applied to IC_{402} which, in turn, develops two square wave signals (the data-slice level control, or DSLC, and the preference pulse, or PREF). DSLC and PREF are combined within IC_{601} and produce an error voltage that becomes the threshold voltage for the EFM comparator.

5.4 LASER SIGNAL PROCESSING

Figure 5.4 shows the basic signal-processing circuits. The EFM signal is applied to pin 47 of IC_{402}, as described in Sec. 5.3. IC_{402} decodes the digital audio signal recovered from the CD and corrects any error that may be in the signal. The EFM signal is first applied to the data-strobe portion of IC_{402}, where audio and control information are separated. The control information is applied to circuits within IC_{402} that perform sync detection, deinterleaving, and the interpolation of audio-data signals.

The signal containing the audio is applied to an error-correction circuit, which

TYPICAL CD PLAYER AND CD-ROM CIRCUITS   5.9

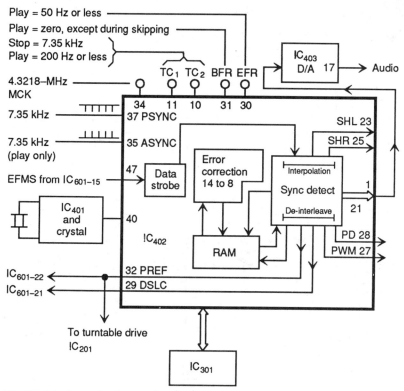

FIGURE 5.4 Laser signal-processing circuits.

converts the 14-bit modulation back to an 8-bit modulation. As discussed in Chap. 1, the audio to be recorded is first converted (during CD manufacture) to an 8-bit digital format and to a 14-bit format for recording on the CD. The restoration process in the CD player is done by shifting the digital data bits into and out of a RAM. The audio data bits are then shifted through the RAM circuit into the circuits where deinterleaving (Chap. 1) and interpolation are performed.

Interpolation is a process of averaging the recovered digital audio signal, where a digit or audio sample may be lost. An average value that falls between the two last-known points or values is generated by the interpolation circuits. From a practical troubleshooting standpoint, interpolation and deinterleaving, sync detection, and modulation restoration are all performed automatically by IC_{402}, as discussed in Chap. 8.

5.5 AUDIO CIRCUITS

Figure 5.5 shows the audio circuits. Note that the right-channel circuits are stressed. Audio output from pin 17 of IC_{403} (Sec. 5.4) is amplified by IC_{501} and

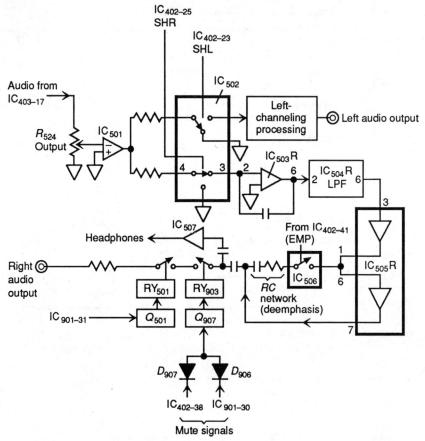

5.5 Audio circuits.

multiplexed into right and left channels by sample/hold (S/H) circuits within IC_{502}, under control of IC_{402}.

Note that the audio output from pin 17 of IC_{403} is passed through front-panel output control R_{524}. Also note that the audio signal is still in a "serial" left, right, left, right format and must be converted to conventional "parallel" stereo audio. The audio from IC_{403} also contains a certain amount of digital noise, which must be filtered out to produce a high-quality signal. This is done by the audio-output circuits.

5.5.1 Serial-to-Stereo Conversion

The SHR and SHL signals generated by IC_{402} (Sec. 5.4) are applied to pins 9 and 11 of IC_{502}. SHR and SHL close the proper switch at the correct time to route the left-audio information through the left-channel processing circuits and the right-audio information to the right channel. When one switch in IC_{502} is closed, the other switch is connected to ground, thus preventing any noise from passing to the processing circuits.

5.5.2 Filtering

The right-channel audio exits IC_{502} at pin 3 and is applied to IC_{503R}. The capacitor between the input (pin 2) and output (pin 6) of IC_{503R} removes much of the digital noise present at pin 3 of IC_{502}. The audio is then applied to pin 2 of IC_{504R}, and "analog" low-pass filter (LPF). The audio exits IC_{504R} at pin 6 and is applied to pin 3 of IC_{505R}. The first stage of amplification within IC_{505R} occurs between pins 1 and 3. The audio reenters IC_{505R} at pin 6, is amplified once again, and exits at pin 7.

An *RC* network is connected across pins 6 and 7 of IC_{505R} for deemphasis of the high-frequency signals. The time constant is cut in and out of the circuit by a switch in IC_{506}. The switch is controlled by the emphasis (EMP) signal at pin 41 of IC_{402}. Deemphasis is only required on CDs that have preemphasis during the record process. IC_{402} recognizes a CD with preemphasis by a "flag" signal recorded on the CD. When the flag is present, IC_{402} switches in the deemphasis *RC* network.

5.5.3 Output Control

The audio at pin 7 of IC_{505} is coupled to the rear-panel output jacks through two relays. Relay RY_{903} is an internal muting relay, operated by mute signals from IC_{901} and IC_{402}. Relay RY_{501} is the audio-bus relay operated by signals from IC_{901}. The front-panel headphone jacks receive audio ahead of the relays through IC_{507}.

5.6 AUTOFOCUS

Figure 5.6 shows the principles and circuits involved in CD player autofocus. As discussed in Chap. 1, a certain amount of eccentricities may occur during the manufacturing process, even though CDs are manufactured to very tight tolerances. To compensate for this condition, a method of automatically focusing the laser beam on the CD surface is incorporated into all CD players.

5.6.1 Autofocus Principles

Compare the following circuit details with the autofocus principles discussed in Chap. 1. As shown in Fig. 5.6*a*, the laser beam reflected from the CD surface passes through an objective focus lens. The reflected beam then passes through coupling and cylindrical lenses. As discussed in Sec. 5.3, the cylindrical lens causes the beam to become elongated if the beam is not properly focused.

After the beam passes through the cylindrical lens, it is focused onto a photosensor (also called an *optical detector, photo transistor*, etc.). The detector is divided into four zones. If the lens is focused correctly on the CD surface, the

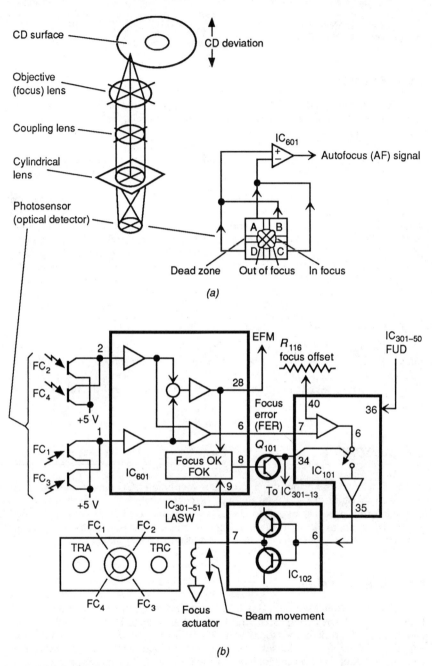

FIGURE 5.6 Autofocus circuits.

reflected beam is perfectly round and generates equal signals from zones A, B, C, and D. When these equal signals are applied to a differential amplifier within IC_{601} (Fig. 5.6b), a zero autofocus signal is generated.

Now assume that the CD deviates in a downward, or negative, direction. This causes the beam striking the detector to become slightly elongated because of the cylindrical lens. With an elongated beam, more light is applied to zones B and D, causing the noninverting input of IC_{601} to go high. As a result, the output of IC_{601} goes positive, repositioning the objective lens as necessary to refocus the beam and get an equal output from all zones of the detector.

If the CD deviates in an upward, or positive, direction, zones A and C produce a higher voltage than zones B and D. This raises the input voltage to the inverting input of IC_{601}. With the inverting input high, the autofocus voltage from IC_{601} goes in a negative direction, refocusing the objective lens to get equal outputs from all four zones of the detector.

5.6.2 Autofocus Circuit

Note that the autofocus sensors, or detectors, shown in Fig. 5.6 are also used to recover the main audio signal (the EFM signal), as discussed in Sec. 5.3. The four zones in Fig. 5.6a are represented by four separate phototransistors FC_1 through FC_4 in Fig. 5.6b. The sensors are transistors that turn on when light strikes the base. The amount of current through the transistors is determined by the amount of light. Sensors FC_2 and FC_4 are placed 180° apart, with the electrical outputs applied to pin 2 of IC_{601}. The outputs of FC_1 and FC_3 are applied to pin 1 of IC_{601}.

When the laser beam is in focus, the outputs from FC_1 through FC_4 are all equal. As a result, the focus error (FER) output from pin 6 of IC_{601} is zero. However, there is always an EFM output from pin 28 of IC_{601}, with the beam in or out of focus. (The EFM signal is removed only when there is no beam or there is no CD and the beam is not reflected back from the CD onto FC_1 through FC_4.)

When the laser beam is out of focus, the outputs from FC_1 through FC_4 are not equal. As a result, there is FER output from pin 6 of IC_{601}. This FER output is applied to the focus actuator through IC_{101} and IC_{102}. The focus actuator moves the focus lens up or down to achieve focus. R_{116} provides an offset adjustment for the FER signal.

When play first begins, the focus actuator receives a focus up-down (FUD) signal from pin 50 of IC_{301} through IC_{101} and IC_{102}. The FUD signals move the focus actuator up and down two or three times as necessary to focus the beam on the CD. Once focus is obtained, a focus OK (FOK) signal is generated by IC_{601} and applied to both IC_{301} and IC_{101} through Q_{101}. Note that IC_{601} does not produce FOK signals unless there is a laser switch (LASW) signal at pin 9, as well as EFM signals at pin 28.

If an FOK signal is not received after two or three tries, IC_{301} shuts the system down and play stops (the turntable stops and the pickup moves to the inner limit). Note that if there is no CD in place, there is no EFM signal and thus no FOK signal. In this way, the FOK function also serves as a CD detector.

5.7 LASER TRACKING

Figure 5.7 shows the principles and circuits involved in CD player tracking of the pits (audio information) on a CD. As discussed in Chap. 1, tracking is done by passing the laser beam through a defracting lens, which separates the main beam into three separate beams that are slightly staggered. The TRA and TRC beam spots are used to generate the tracking-error signal, or TER (similarly to FER). The main beam spot is used to provide both the focus signal and main audio signal.

5.7.1 Tracking Principles

Compare the following circuit details with the tracking principles discussed in Chap. 1. As shown in Fig. 5.7a, TRA and TRC are shifted slightly off center of the main spot (in opposite directions). The TRA and TRC beams are reflected back from the CD track into the prism mirror and onto the tracking photosensors (on the same block as the focus and audio sensors but located on either side of focus and audio).

The outputs of the tracking sensors are applied to a differential amplifier, which generates the correction signal. For example, if the main spot shifts to the left of the desired track, TRC moves completely off the pits, thus reflecting very little light. At the same time TRA moves over the pits, thus reflecting more light. This causes more laser light to fall on TRA, generating a higher TRA output and causing the inverting input of the differential amplifier to go high. This allows the output of the differential amplifier to generate a positive voltage, causing the tracking actuator (for all three beams) to move to the right (and thus restore proper tracking).

Now assume that the main spot shifts to the right of the pits. The TRA output goes low while the TRC output goes high. Since TRC is directly over the pits, more of the laser beam is reflected onto the TRC sensor. This TRC output is coupled into the noninverting input of the differential amplifier, causing the amplifier to generate a negative output, moving the tracking actuator (and beams) to the left (to again restore proper tracking).

5.7.2 Tracking Circuits

As shown in Fig. 5.7b, the outputs from TRA and TRC apply signals to pins 2 and 6 of IC_{603}. The TRA and TRC signals are coupled to pins 1 and 7, respectively, through amplifiers within IC_{603}. When the laser beams track the CD correctly, the outputs at pins 1 and 7 are equal.

The TRA output is coupled through tracking offset R_{603} to pin 4 of IC_{601}. R_{603} provides a fine adjustment so that the output at pin 5 of IC_{601} is zero when the CD is tracking properly.

The TRC signal from pin 7 of IC_{603} is passed through delay line CP_{603} and CP_{604} before entering IC_{601} at pin 3. CP_{603} and CP_{604} delay the TRC signal as

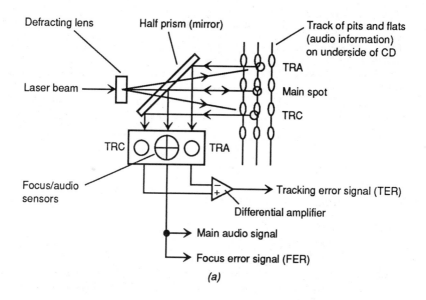
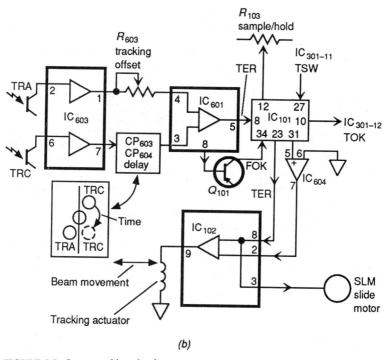

FIGURE 5.7 Laser-tracking circuits.

necessary so that TRC arrives at IC_{601} simultaneously with TRA. The time delay is necessary because the tracking circuits require that the TRA and TRC beams analyze the same point on the CD, even though the beams are separated by the optical system.

The TER (if any) at pin 5 of IC_{601} is applied through IC_{101} and IC_{102} to the tracking actuator coil and causes the actuator to move the beams right and/or left as necessary to produce proper tracking. Note that IC_{101} requires various signals (FOK, TSW, etc.) to process the TER signals properly. Also note that S/H R_{103}, discussed in Sec. 5.2.2, used by the SLM, has an effect on the tracking circuits. These factors are discussed in Chap. 8.

5.8 TURNTABLE MOTOR CIRCUITS

Figure 5.8 shows the CD turntable motor circuits. As discussed in Chap. 1, the CD turntable is rotated at a variable speed. This keeps the rate at which the track moves (in relation to the pickup) constant. (The track is the series of pits on the CD that represent audio information.) The speed variations are necessary since there is less information on the tracks near the inside of the CD (start) than near the outside (stop or end).

Most CD players use some form of unitorque motor with Hall-effect elements to get the variable speed. This is similar to the speed-control circuits for LP turntables. (If you are not familiar with LP turntables, read *Lenk's Audio Handbook* immediately.) Of course, with LP turntables, you want constant speed or constant angular velocity (CAV), instead of the constant linear velocity (CLV) required for CD players. Typically, the CD speed varies from about 480 rpm (inside) to 210 rpm (outside) so as to maintain a CLV of about 1.25 to 1.3 m/s.

In the circuit in Fig. 5.8, the Hall-effect outputs are fed through IC_{201} to the motor drive windings (A and B) and thus maintain the desired speed. The Hall-effect elements are also fed currents through IC_{201} (from controller IC_{402} under the direction of IC_{301}) to vary the speed at the desired rate. CLV circuits within IC_{402} monitor the EFM signal to determine the rate at which information is passing and then produce the necessary signals to maintain the desired speed. R_{201} sets the gain of the Hall-effect signals in IC_{201} and thus sets the motor speed.

Operation of the motor control can be divided into two phases: when power is first applied (sometimes called the start-servo phase) and when the motor reaches the desired speed (the regular-servo phase).

When power is first applied, the motor runs freely, DMSW and CLVH signals (applied to IC_{402} and IC_{201} from IC_{301}) are low, and the ROT signal (from IC_{301} to IC_{402}) is high. Under these conditions, the motor begins to accelerate and turn at a constant velocity. IC_{402} then produces essentially similar outputs at PWM, PREF, and PD.

After a free-run period (set by IC_{301}), ROT goes low and the motor starts to accelerate. The EFM signal is read by IC_{402} and compared to a reference. The difference between the reference and the EFM is the PWM output from IC_{402}.

During the acceleration portion of the start-up period, the PWM duty cycle (which varies above and below 50 percent, as determined by motor speed) is

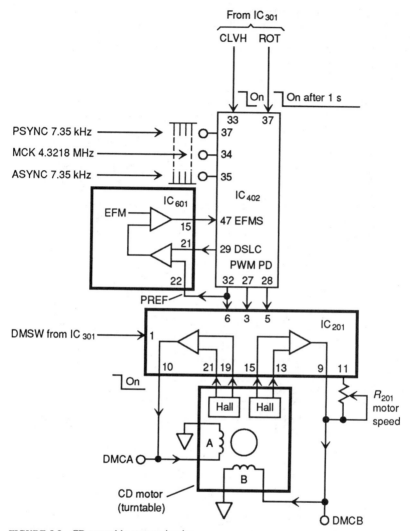

FIGURE 5.8 CD turntable motor circuits.

compared with the PREF signal (which is a fixed duty cycle). The result of this comparison is applied to the motor circuits to control speed.

When the motor reaches the desired speed (the regular-servo phase), the PWM signal has a 50 percent duty cycle and the pickup reads the CD data at a CLV.

The CLV condition is maintained within a ±1 percent accuracy by the PD pulse from IC_{402}. The duty cycle of the PD pulse is set by comparison of the EFM signal to a reference within IC_{402}. In turn, the PD signal is compared with the output from the PWM and PREF comparator. The result of this comparison is applied to the motor and maintains the 1 percent accuracy.

5.9 ANTISHOCK CIRCUITS

In addition to the error-detection circuits, some CD players include antishock circuits to minimize the effects of severe shock and/or vibration. Figure 5.9 shows typical antishock circuits. The related waveforms are shown in Fig. 5.10. The antishock function is controlled by a switch (usually on the rear panel) and is on only when the player is in a location with large amounts of vibration or when the disc has excessive eccentricity. Sound skipping can occur in players operated under either of these conditions.

With the antishock switch on, the antishock circuits compensate for vibration and disc eccentricity. However, antishock circuits can actually increase sound skipping in rare cases (such as when the disc is badly scratched). So, *when troubleshooting a sound-skipping problem, always check the antishock switch setting.*

Most antishock circuits operate by gating and holding the audio information (Fig. 5.4) at the previous level for a specific time interval (until the shock is over). The circuits also operate when there is a scratch on the disc. In fact, when there

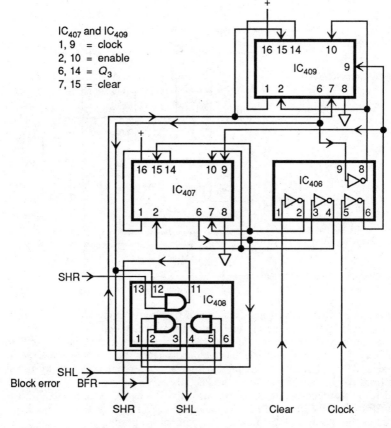

FIGURE 5.9 Antishock circuits.

TYPICAL CD PLAYER AND CD-ROM CIRCUITS 5.19

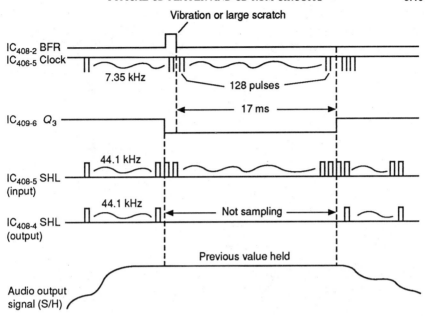

FIGURE 5.10 Antishock circuit waveforms.

are many scratches, the antishock circuit can produce more skipping than normal. This is because the antishock circuit operates from a block error (BFR) signal generated by the player controller (such as the BFR signal at pin 31 of IC_{402} in Fig. 5.4). The BRF signal occurs only when the controller detects a large block error (resulting from shock or defects in the disc).

When the antishock circuits are turned on by the antishock switch, and a BFR signal is present, the audio information is held at the previous level for a time determined by IC_{406} through IC_{409}, which are combined as a counter circuit. Typically, the time is about 17 ms after detection of skipping (after the BRF signal is present). Most shock noises occur during this period.

When troubleshooting antishock circuits, remember that a scratch can also produce a BFR signal. In turn, this triggers the antishock circuits into holding the audio information for the 17 ms, even though the scratch may produce a block error of much shorter duration. This is because the holding period is set by the counter circuits and not by the size of the scratch (or the amount of block error).

5.10 JUMP CIRCUITS

Most CD players are capable of forward and/or reverse jumps of tracks from any point on the disc. Figure 5.11 shows typical jump circuits. The related waveforms are shown in Fig. 5.12. The jumps are initiated by JPF (jump forward) and JPR (jump reverse) signals from the system-control microprocessor (such as IC_{301} in Fig. 5.1).

When a JPF signal is sent by IC_{301}, the pulse is applied to the set (S) input of

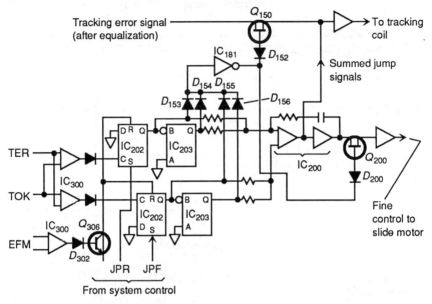

FIGURE 5.11 Jump forward and jump reverse circuits.

a D-type flip-flop (FF) in IC_{202}, as shown in Fig. 5.11. The Q output of the FF goes high. This disables Q_{200} and Q_{150} (through D_{155}, IC_{181}, D_{152}, and D_{200}) and shuts off the TER signal to the tracking coil and the slide motor. (In most CD players, the TER signal provides control of the tracking coil as well as a *fine control* for the slide motor.)

During jump, the slide motor receives no fine-control signal, and the tracking coil is controlled by the outputs of the circuits shown in Fig. 5.11 (which are summed at the output of IC_{200}). This output drives the tracking coil toward the next track.

To notify the jump circuits when the pickup is centered on the next track, the TER and tracking OK (TOK) signals are compared by IC_{300}. Figure 5.12 shows the relationships among EFM, TER, TOK, and JPF IC_{300} input-output, and the IC_{200} jump-pulse outputs. As shown, the TER S-curve goes high and the TOK pulse goes low when the tracking beams leave the present track. TER and TOK are compared by IC_{300} and produce a low at the IC_{300} output.

As the tracking beams become centered between tracks, the TER signal passes zero (going in the negative direction). The TER signal continues in a negative direction as the tracking beams move to the next track. As soon as the TER signal drops below 0 V, the output of IC_{300} goes high.

The low-to-high transition at the output of IC_{300} changes the state of the IC_{202} JPF FF and triggers a one-shot FF within IC_{203}. This pulls the output of IC_{200} negative for a fixed time interval, as shown in Fig. 5.12. The negative-going one-shot pulse acts as a "brake" to counter the inertia of the radial-tracking mechanism, causing the mechanism to stop when aligned with the next track.

The EFM signal is applied through IC_{300}, D_{302}, and Q_{306} to the reset inputs of the IC_{202} FFs. This restores operation to normal when the jump is complete.

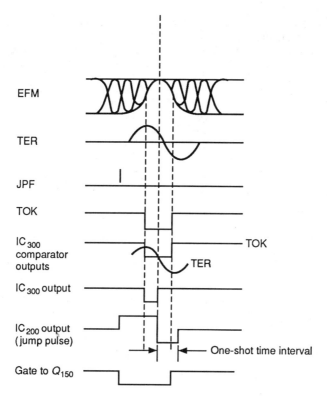

FIGURE 5.12 Waveforms associated with jump forward and jump reverse.

Note that JPR operates the same way as JPF, except for the polarity of the output pulse at IC_{200}.

5.11 INTRODUCTION TO CD-ROM

The CD-ROM technology was derived from the audio CD players discussed thus far in this chapter. As a result, many of the circuits are identical, so we concentrate primarily on the differences between audio CD and CD-ROM in this section.

Instead of audio, computer information is stored on a CD-ROM disc as variable-length pits in a continuous spiral track. The pits are burned into a master disc by a laser beam and then used to press holes and flat surfaces into each copy disc.

As in the case of audio, a CD-ROM disc is read by a scanning laser beam that reflects from the silvered areas. The information is stored in the continuous spiral track and is read by the drive at CLV (the disc spins at a speed inversely proportional to the radius being read). Since each section of data must be read in the same amount of time, the disc spins faster on inner tracks and slower on the outer tracks (as does an audio CD).

A CD-ROM *drive* is similar to an audio CD *player* except that the CD-ROM is designed to handle computer data instead of audio. The CD-ROM drive is operated through a computer and software, rather than by direction operation using controls (front panel or remote). A CD-ROM is usually connected to a computer to deliver large amounts of data in a usable fashion.

CD-ROM drives, which include audio output (with headphones) or line-level outputs, can also provide audio in selectable segments or can be used to play standard audio CDs under host-computer control. Of course, software for playing audio CDs is needed to play standard CDs on CD-ROM drives.

Since this is primarily a troubleshooting and service book, we are concerned with circuits rather than programming, so we do not discuss software. The troubleshooting approach here assumes that the computer and drive system operated properly at one time and that there are no software or program problems.

CD-ROM is used for very large databases and for reference materials, such as medical, legal, or government libraries. CD-ROM products include maps, catalogs, training courses, and educational books. Any data that can be represented digitally can be stored on CD-ROM discs. This includes text, music, pictures, and computer graphics.

One CD-ROM disc can store about 552 Mbytes of digital data when 60 min of disc space is used for mode-1 data. When mode-2 is used for data storage, 60 min holds about 630 Mbytes. It takes the equivalent of a 1,533,360-kbyte DD/DD 5.25-in floppy disc to store the same amount of information. Approximately 270,000 typed pages of text can be reduced to just one CD-ROM disc. At the present time, CD-ROM discs and drives are read-only.

5.11.1 Major Differences between CD-ROM and CD Audio

The following is a summary of the major differences between CD-ROM drives and audio CD players:

1. *Minimal user controls and displays:* A typical CD-ROM drive will have a power switch-LED and possibly some additional LEDs to show what functions are being performed in the CD-ROM system microprocessor (such as disc-tray open and close, "busy" functions where the computer and drive are accessing data from the disc, etc.). A drive-select switch is also found on some CD-ROMs, particularly when more than one CD-ROM is used or when other drives (floppy, hard, etc.) are used. This makes it possible for the computer to select drives in a given order or to omit drives, as necessary.

2. *A Bidirectional data and communications port*, and control lines, between the host computer and the CD-ROM system microprocessor. Figure 5.13 shows the basic connections. Electrical interface connections between the CD-ROM drive and host computer are discussed further in Sec. 5.11.5.

3. *There are five layers of internal error correction* for audio data and CD-ROM data, compared to four layers for a typical audio-only CD player.

4. There are minimal audiophile circuit features (or often none at all) in a CD-ROM drive.

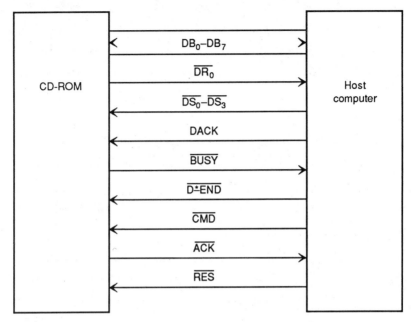

FIGURE 5.13 Basic connections between CD-ROM and host computer.

5. A CD-ROM drive provides much greater programmability through use of the host-computer resources and programming.
6. A CD-ROM drive has the ability to return Q-code, drive-status, or CD-ROM data to the host computer. Figure 5.14 shows a comparison of CD-ROM and audio CD formats, both in sector format and Q-code format. Basic CD-ROM formats are discussed further in Sec. 5.11.4.
7. A CD-ROM drive has the ability to transfer 153,600 bytes of data per second for over 60 min from a CD-ROM disc.
8. A CD-ROM provides very high-speed access to the inner and outer tracks of a disc. Some CD-ROM drives have a front-panel control (often called FG In or some similar term) that drives the slider at high speed to inner or outer tracks (similar to the fast-forward or fast-reverse functions of an audio CD player).
9. A CD-ROM drive converts serial data to 8-bit parallel data using a CD-ROM conversion IC and a RAM.
10. Most CD-ROM drives use direct-drive (DD) disc motors.
11. *The major circuit differences found in a CD-ROM* are shown in Fig. 5.15. These include an IC used for error correction and data conversion (IC_{302}), additional RAM (IC_{303}) for the CD-ROM data and control circuits, and input-output (I/O) line drivers and interface circuits (IC_{304} and IC_{305}), all of which are discussed further in Sec. 5.11.6. Also note that there are differences in the CD microprocessor IC_{104}, such as an FG In input, as well as outputs to LEDs that show open and close and "busy" or "access" functions.

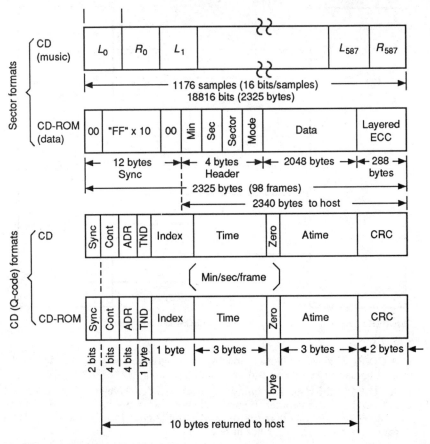

FIGURE 5.14 Comparison of CD-ROM and audio CD formats.

5.11.2 Typical CD-ROM Drive Hookup

This section describes the basic hookup or physical connections between a host computer and CD-ROM drive (internal or external). The pin-by-pin electrical interface is described in Sec. 5.11.5. The following discussion assumes that the host computer is an IBM PC-XT/AT, PS/2 Model 30, or any direct hardware-compatible computer running MS-DOS (Microsoft disk operating system).

External Drive. The hardware installation is fairly simple with external CD-ROM drives. You simply remove the computer cover, plug one of the compatible PC-bus I/F cards (a CDIFI4A, for example) into a vacant 8-bit slot (note that AT-type computers have 16-bit and 8-bit slots). Then connect the I/F cards-compatible external CD-ROM drive (such as the Hitachi CDR-1502S) with a matching cable (such as a CDCBL). No internal switches in the computer need be set.

TYPICAL CD PLAYER AND CD-ROM CIRCUITS

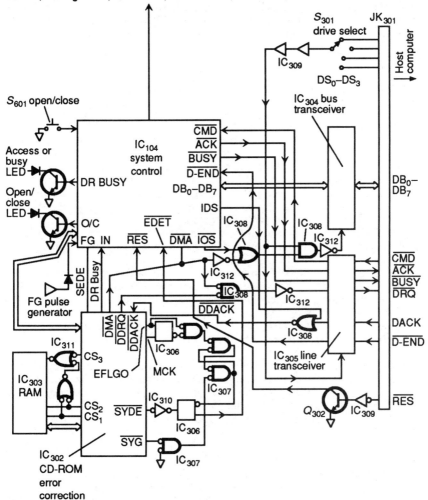

FIGURE 5.15 Major circuit differences between CD-ROM and audio CD players.

Internal Drive. Internal drives (such as the Hitachi CDR-2500 or CDR-3500) require more effort than external drives. First, there must be room inside the computer to install the drive. For example, front-panel clearance and a full-height slot is necessary for the Hitachi CDR-2500. A half-height slot is needed for a Hitachi CDR-3500.

A full-height slot is the same size as a full-height floppy disk drive, while a half-height slot is the same as a half-height floppy. Thus, two half-height devices can be mounted in the same space used by a single full-height floppy disk drive.

Special mounting hardware is required for internal installation. (Mounting kits are available from many computer stores.)

The internal drive requires power from the host computer and generally uses the same power connector as a floppy disk drive. Obviously, if there is no empty power connector from the main power connector, a power connector must be provided.

5.11.3 Recognizing and Accessing the CD-ROM Drive

This section summarizes how a CD-ROM drive is recognized and accessed by a host computer when the drive handler works through MS-DOS.

When the computer is first turned on, the microprocessor jumps to a particular address in memory and starts executing machine-language instructions in ROM. The computer runs various tests and prepares to run some type of user software. Then the computer checks for a disk drive and "boot up" DOS software from a floppy or hard disk. Some computers do not boot to DOS from a disk (when the computer has the disk-handling software in ROM). However, most ROM-based MS-DOS computers still follow the next step.

When the DOS code is loaded, the device-handler code (such as Config.sys in MS-DOS) installs the device name and handler code into the device table. (If you are interested, read the MS-DOS manual for details of this file.)

The computer then looks for a file of autorun software (called Autoexec.bat in MS-DOS) and executes the code found in that batch file. When these programs are finished running, the computer turns control over to the user at the DOS prompt.

After the device handler is installed, the CD-ROM drive is accessed as the *next drive in the system* (except the drive is read-only). Files cannot be saved in a CD-ROM disc or drive.

If the device-handler code for a Hitachi CD-ROM drive is on the boot disk at boot-up, the CD-ROM drive can be recognized by the operating system. Generally, the CD-ROM becomes the next drive after A and B and any other drive handlers installed by the device-handler drive. Typically, the CD-ROM becomes drive C if there is no hard drive installed, or drive D if a 20-Mbyte hard-drive is installed.

5.11.4 CD-ROM Information Layout

This section summarizes the organization of information on a CD-ROM disc. Note that the origination is similar, but not identical, for audio CD and CD-ROM. In both cases, the bit stream has sync signals to allow the bit data to be separated for use in maintaining a constant-speed data stream.

During the disc recording process, data bytes are interleaved (Chap. 2), error-correction information is added, and the data bits are converted from 8 to 14 bits and recorded in a non-return-to-zero format. This is reversed in play by a custom CD decoding chip set and 2-kbyte static RAM. The chip set (IC_{301} through IC_{305}) for a typical CD-ROM provides five layers of internal error correction, compared to the usual four layers used in audio CD players.

Frame Information. As shown in Fig. 5.14, the low-level data bits are set into *bit frames* (similar to a line of video) and *frames* (a group of 98-bit frames, similar to a frame of video). The frame rate is 75 Hz, so one frame occurs in 1/75th of a second. Each frame contains 98 bytes of control information (channels P to W) and 2353 bytes of data.

Control Code Channels. Each bit of control information is considered a *channel*, P, Q, R, S, T, U, V, and W. Channels P and Q are generally used in CD-ROM. Ten bytes (80 bits of Q-channel or Q-code information can be returned to the host computer. Q-code contains information about the type of information on the disc and timing information for positioning. There are Q-code data bits in the table of contents (TOC) area and in the program area.

The TOC of a disc should not be confused with a *directory of files* on the disc. Any directory of files on a disc is contained in individual sectors according to the data format (such as High Sierra, ISO9660, etc.) or custom formats.

The program area of the disc is identified in the Q-code by absolute time in minutes (0–59+), seconds (0–59), and frames (0–74).

CD-ROM discs are usually rated for 60 min of mode-1 data. Audio CD discs are generally considered to have an upper limit of 74 min. The maximum possible upper limit is 99 min for BCD-format data but is limited to 89 min with small-computer (SCSI) command formats.

Frame Data. The 2353 bytes of data contain 1176 words of audio, or 1 sector of CD-ROM data. When the data bits are in a CD-ROM, the bits are passed to the custom CD-ROM controller IC_{302} and the 6-kbyte RAM IC_{303} for serial-to-parallel conversion and to other processing in IC_{301}. After conversion, 4 bytes of header, 2048 bytes of data, and 288 bytes of L-ECC are returned to the host.

The CD-ROM header contains the CD-ROM block in Min, Sec, Block and a data Mode byte. The Mode byte identifies whether the 288 bytes are L-ECC (mode-1) or other (mode-2). The CD-ROM block is not necessarily the same as the Q-code frame number.

5.11.5 CD-ROM Electrical and Data Interface

Figure 5.15 shows a typical electrical and data interface between a CD-ROM and a host computer. Note that the interface can be divided into two sections: (1) the 8-bit parallel data and (2) the control signals. The data bus transmits all information (the *data read* from the CD-ROM, the *control-command code*, and the *status information* of the CD-ROM system). The control signals control transmission of this information.

Interface Lines and Buses. The following describes the function of each interface line and bus between the CD-ROM and host computer (Figs. 5.13 and 5.15):

> *DB0-7 (data bus):* DB is a bidirectional 8-bit parallel bus. Data bits from the CD-ROM, control-command code to the CD-ROM, and status information of the CD-ROM system are transmitted through this 8-bit data bus. All 8 bits are passed to and from the CD-ROM system-control microprocessor IC_{104} through interface connector JK_{301} and bus transceiver IC_{304}.
>
> System-control processor IC_{104} is connected to the CD-ROM error-correction processor IC_{302} through a bus and control lines. Note that CD-ROM processor IC_{302} is not to be confused with the CD signal processor IC_{301}, which has similar functions to the signal processors of audio CD players. Both processor

IC_{302} and RAM IC_{303} are unique to CD-ROM players and provide the five layers of error correction necessary for use with a computer.

$\overline{DS}_0$ *through* $\overline{DS}_3$ *(drive select):* $\overline{DS}$ is the signal input which enables the selected CD-ROM drive. The CD-ROM drive is selected when one of $\overline{DS}_0$ through $\overline{DS}_3$ goes low. When $\overline{DS}_0$ through $\overline{DS}_3$ are high, all signal lines are in the high-impedance state. The 4 DS bits are used to control both IC_{304} and IC_{305} through the front-panel drive-select switch S_{301}, IC_{309}, IC_{308}, and IC_{312}.

$\overline{DRQ}$ *(data request):* $\overline{DRQ}$ is the signal output which controls the data communications. When $\overline{DRQ}$ is active (low), the CD-ROM data on the 8-bit data bus is valid (and only the $\overline{DRQ}$ is low). $\overline{DRQ}$ shakes hands with $\overline{DACK}$ (data acknowledge) and transfers the data from the CD-ROM to the host computer. $\overline{DRQ}$ is a combination of $\overline{DMA}$ from IC_{104} and $\overline{DDRQ}$ from IC_{302}, and it is applied to the host computer through IC_{308}, IC_{305}, IC_{312}, and JK_{301}.

DACK (data acknowledge): DACK is the signal input which controls the data communication. When the host computer has read data from the CD-ROM system, it makes DACK active (high). Then the CD-ROM makes $\overline{DRQ}$ inactive (high) after confirming that DACK will become active (high). The CD-ROM outputs the next data on the data bus and makes $\overline{DRQ}$ active (low) when DACK becomes inactive. DACK is applied to IC_{302} (as $\overline{DDACK}$) through JK_{301}, IC_{305}, and IC_{308}.

$\overline{BUSY}$ *(bus busy):* $\overline{BUSY}$ is the signal output which informs the condition (accessing or not accessing) of the data bus. While IC_{104} is in the data-transmission mode, a command code cannot be accepted from the host computer. During the data-transmission mode, IC_{104} makes $\overline{BUSY}$ active (low). When IC_{104} receives a $\overline{D\text{-}END}$ signal from the host computer, IC_{104} makes $\overline{BUSY}$ inactive (high).

$\overline{D\text{-}END}$ *(data transfer end):* $\overline{D\text{-}END}$ is the signal input which indicates the end of the data-transmission mode. IC_{104} changes from the data-transmission mode to the command mode and makes $\overline{BUSY}$ inactive (high) after receipt of a $\overline{D\text{-}END}$ signal from the host computer. $\overline{D\text{-}END}$ is applied to IC_{104} through JK_{301} and IC_{305}.

$\overline{CMD}$ *(command pulse):* $\overline{CMD}$ is the signal input which controls transmission of command codes and status information. IC_{104} transfers status information, or receives command codes, by shaking hands with the host computer through $\overline{CMD}$ and $\overline{ACK}$ signals. When $\overline{BUSY}$ is active, IC_{104} does not accept a $\overline{CMD}$ signal. $\overline{CMD}$ is applied to IC_{104} through JK_{301} and IC_{305}.

$\overline{ACK}$ *(acknowledge):* $\overline{ACK}$ is the signal output which controls the transmission of command codes and status information. $\overline{ACK}$ becomes active (low) after receipt of command codes or when status information is being output. $\overline{ACK}$ shakes hands with the $\overline{CMD}$ signal, and it is applied to the host computer through IC_{305} and JK_{301}.

$\overline{RES}$ *(reset):* $\overline{RES}$ is input from the host computer. If $\overline{RES}$ is active (low), IC_{104} and the CD-ROM are reset. $\overline{RES}$ is applied to IC_{104} through JK_{301} and IC_{305}.

CHAPTER 6
TYPICAL CDV PLAYER CIRCUITS

This chapter describes the theory of operation for a number of CDV player circuits. The circuits described here are all the ones that are associated with mechanical components, power supply, mute functions, spindle-motor start-up, system control, system start-up, servo, focus, tracking, tilt, signal processing, timebase correction, spindle servo, noise reduction, on-screen display, memory video, analog audio, and digital audio. The mechanical functions are discussed in Chap. 7.

After studying the circuits found in this chapter, you should have no difficulty in understanding the schematic and block diagrams of similar CDV players. This understanding is essential for logical troubleshooting and repair, no matter what type of electronic equipment is involved. No attempt is made to duplicate the full schematics for all circuits. Such schematics are found in the service literature for the particular CDV player.

Instead of a full schematic, the circuit descriptions are supplemented with partial schematics and block diagrams that show important areas such as signal-flow paths, input-output, adjustment controls, test points, and power-source connections (the areas most important in service and troubleshooting). By reducing the schematics to these areas, the circuits are easier to understand, and you will be able to relate circuit operation to the corresponding circuit of the player you are servicing.

6.1 RELATIONSHIP OF CDV PLAYER CIRCUITS

Figure 6.1 shows the relationship of the player circuits. The line supply is applied to the power transformer through the power switch and fuses in the usual manner. The power transformer voltages are distributed to the player circuits through regulators, as described in Sec. 6.2. The following paragraphs describe the major functions of player circuits. Full circuit details are discussed in Secs. 6.2 through 6.17.

6.1.1 Start-Up and Loading

When power is first applied, the player goes into a start-up mode to check for the presence of a disc. Control of the start-up operation is provided by the system

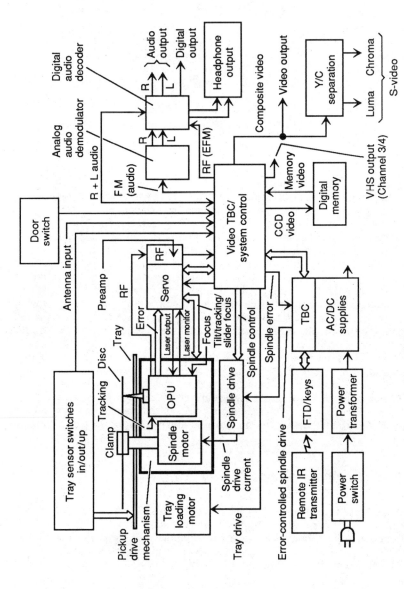

FIGURE 6.1 Relationship of CDV player circuits.

control circuits (Sec. 6.5). First, the tray-loading motor is instructed to turn on and perform clamping, which includes moving the tray and the clamper assembly to position a disc on the spindle. (This is similar to the clamping functions discussed in Chaps. 5 and 7.)

The clamping operation is determined by the position of the tray, indicated by the in, out, and up sensor switches. If the tray is in the up position, the loading tray and clamper assembly move down to clamp a disc (if present) onto the spindle. At the same time, the optical pickup unit (OPU) is moved by the slider drive from the park position to the LD-search position (a position to check for the presence of a 12- or 8-in Laservision disc).

The LD-search position is outside the diameter of a 5-in CD or CDV, about 2½ in from the center of the spindle. A second clamping operation is performed as the OPU moves out to the correct LD-search position. Only one clamping operation is performed if the tray is in the down position when the player is turned on.

6.1.2 Focus

After the OPU reaches the LD-search position and the clamping sequence is completed, the laser diode turns on, and the focus-search sequence is executed (through the servo-control bus) by commands from system control. If focus is found (indicating that a CDV disc is in place), the laser turns off and system control receives data from the servo to indicate the presence of a CDV disc. The OPU then returns to the park position and the player is placed in the stop mode, awaiting instructions from the front-panel keyboard and/or infrared (IR) remote unit.

If focus is not found (indicating that there is no CDV disc in place), the OPU is instructed to move to a CD-search position (about 1¼ in from the center of the spindle). The focus search sequence is then repeated to search for a 3- or 5-in CD or a 5-in CDV (CDV Single or Gold CD).

If a disc is present, the spindle drive receives a command from system control (through the spindle-control line) to turn the spindle motor on. The OPU then moves to the correct position to read the table of contents (TOC) on the disc. After reading the TOC, the laser diode turns off, and the OPU returns to the park position. The TOC is decoded by system control, and information regarding the disc (playing time, number of tracks, etc.) is displayed on the front-panel fluorescent tube display (FTD) and on the TV screen. The player remains in the stop mode until play is activated by the keyboard or remote unit.

If there is no disc present, the OPU returns to the park position and the player is placed in the stop mode, awaiting the command to open the tray so that a disc can be inserted. Keep this in mind when troubleshooting a "the player turns on and off again but will not play" symptom. Make certain that there is some kind of disc in place.

6.1.3 Keyboard and Remote Commands

Commands from the keyboard or remote control are applied to system control through circuits on the power-supply board. Likewise, control and data information to the FTD is applied through the power-supply circuits. Command information includes the RC_5 code (Philips remote control transmitter code 5) sent to system control. The RC_5 code is also sent out to an external device (such as an amplifier) or received from an external device through an RC_5 in-out jack. The system-control circuit also receives signals from the power supply, the door switch, and the in, out, and up switches.

6.1.4 Spindle Drive

The system control circuits provide the spindle drive with a spindle-control signal and provide the power supply with a spindle-error signal. This error signal is developed by the time-base correction (TBC) circuit. The error-controlled spindle-drive signal is then provided to the spindle-drive circuits.

The spindle drive provides spindle-drive current to the spindle-motor coils. When the disc spins, the optical pickup reads the disc and develops both RF and error signals. The error signals are sent to the servo to develop radial and focus error signals. In turn, the radial and focus error signals are fed back to the pickup drive mechanism, keeping the pickup in focus and moving over the disc at the correct speed. The RF signals are processed by system control.

6.1.5 RF Signal Processing

The system-control circuits extract three components from the RF signals recorded on the disc: video FM, analog audio FM, and digital audio HF. The analog audio FM is sent to the analog audio circuit for demodulation. The demodulated right and left audio signals are sent to the digital audio circuits. The digital audio HF (when present) is also sent to the digital audio circuits.

If the disc is recorded in analog only, the analog audio is automatically selected and routed to the audio-output jacks. If the disc is encoded with digital audio, the digital audio is selected and routed to the audio-output jacks. If the disc contains both analog audio and digital audio, the audio-output source can be selected by the user.

The selected audio signals (right and left) are also sent to the headphone-output circuits. Likewise, the audio signals are combined and applied to the RF modulator.

6.1.6 Composite Video Signal

The video FM signal (extracted from the RF signals) is demodulated by system control to provide the composite video signal. The composite signal is applied to digital-memory circuits to provide digital special effects (memory, still, strobe, etc.). The digitally processed signal (when used) is then returned to the system-control circuits.

The composite video signal (with or without digital processing for special effects) is sent to the video-output jack and to Y/C separation circuits. When used, the Y/C separation circuits provide separate chroma and luma to the S-video jack (for use with TV monitors capable of S-VHS operation).

The system-control circuits also provide video and audio (L + R) signals modulated on channel 3 or 4 through the VHF output jack. As with most VCRs, the

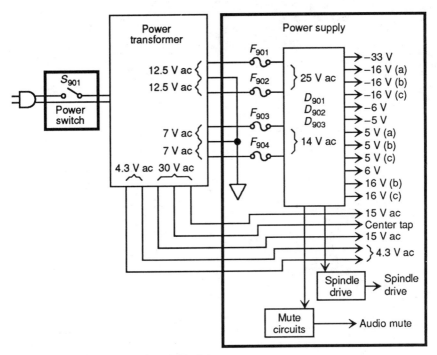

FIGURE 6.2 Power-supply circuits in block form.

antenna or cable signal may also be routed to the VHF output jack if selected (when the player is not in use).

6.2 Power-Supply Circuits

Figure 6.2 shows the power-supply circuits in block form. Full circuit details are discussed further in Chap. 9. Note that the power supply for our CDV player is contained on the following three circuit boards: power-switch, power-transformer, and power-supply boards. In addition to the power-supply circuits, the power-supply board contains the mute circuits (Sec. 6.3) and the spindle-drive circuit (Sec. 6.4).

6.3 MUTE CIRCUIT

Figure 6.3 shows the mute circuit and related timing charts. The mute circuit mutes the audio so that a "pop" sound is not heard in the speakers when the player is turned on or off. The voltages shown in Fig. 6.3 are present after the player is turned on. In this configuration (after initial turnon) both Q_{909} and Q_{910} are off.

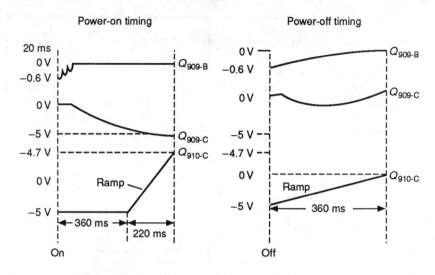

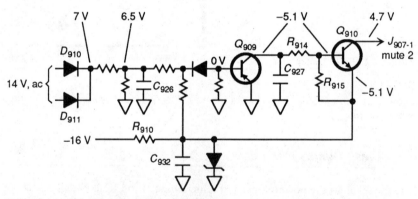

FIGURE 6.3 Mute circuit and related timing chart.

When power is initially applied, 14 V is applied across rectifiers D_{910} and D_{911}, producing about 7 V. At the same time, -16 V is applied to the mute circuit through R_{910}. When the player is turned on, Q_{909} is forward-biased for about 20 ms. During the time that Q_{909} is on, current flows through R_{915} and R_{914}, forward-biasing Q_{910}. After Q_{909} turns off (after 20 ms), C_{927} is allowed to charge, permitting current to continue flowing through R_{914} and R_{915}.

While C_{927} is charging, Q_{910} is in saturation for about 360 ms and applies about -5 V to pin 1 of J_{907}. This -5 V at J_{907-1} is the mute-2 signal. C_{927} continues to charge for another 220 ms until the -5 V is reached. The mute-2 signal rises up to about 4.7 V (forming a ramp signal, as shown), permitting the audio to have a soft turn on from mute.

When the player is turned off, the base of Q_{909} goes momentarily low (about -0.6 V). The positive 6.5 V across C_{926} discharges quickly, applying the negative potential from charged C_{932} to the base of Q_{909}. When Q_{909} is on, Q_{910} is forward-biased to apply the mute-2 signal to pin 1 of J_{907}. It takes about 2.5 s for C_{932} to discharge and turn Q_{909} and Q_{910} off.

6.4 SPINDLE-MOTOR START-UP

Figure 6.4 shows the spindle-motor start-up circuits. Figure 6.5 shows the spindle-motor drive circuits and related timing charts. The spindle servo is discussed in Sec. 6.13.

To start the spindle motor, the spindle circuits receive two control signals from system control: the run (start-stop) signal and the forward-reverse (F/R) signal. A frequency generator (FG) signal, developed by the spindle-motor IC QD_{01}, is returned to system control (Sec. 6.5) through pin 3 of JD_{02}.

As shown in Fig. 6.4, QD_{01} receives motor-rotation information from the Hall sensors (HA, HB, and HC) located on the spindle-motor board (adjacent to the spindle). When QD_{01} receives a high (5 V) at the run input (pin 9), the spindle motor begins to rotate. As soon as the motor rotates, the FG signal developed by QD_{01} is sent to system control, thus monitoring the spindle-motor speed.

When play is activated, the F/R signal (pin 8) goes low (0 V) to start the clock-

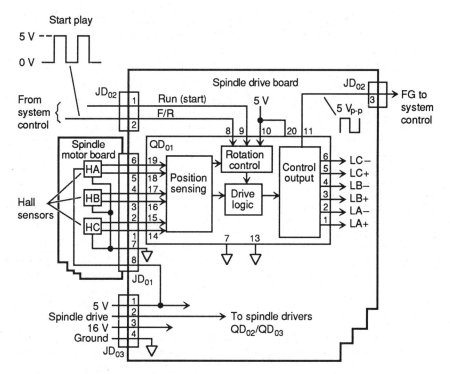

FIGURE 6.4 Spindle-motor start-up circuits.

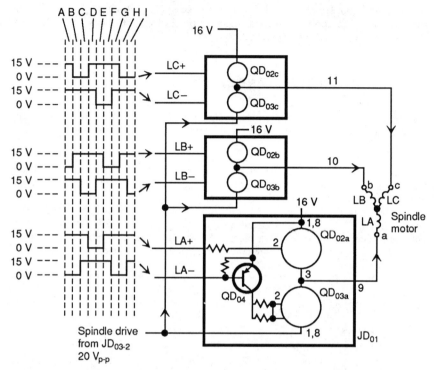

FIGURE 6.5 Spindle-motor drive circuits.

wise, or forward, rotation of the spindle motor. When the motor reaches the correct speed (spindle-lock condition), the F/R signal toggles from low to high several times at various intervals to prevent runaway of the spindle-motor speed. When the motor settles to the correct speed, the F/R signal goes low and remains low until the stop mode is activated.

When stop is activated, QD_{01} receives a high at the F/R input to stop the motor. Actually, this places QD_{01} in the reverse (CCW) mode to brake the spindle-motor rotation. (On our player, the brake input at pin 10 of QD_{01} is not used. Instead, pin 10 is connected to 5 V.) When the motor stops, the run signal goes low and the motor is in the stop mode. The F/R signal remains high until play is again activated.

As shown in Fig. 6.5, QD_{01} controls the switching of the spindle-driver transistors QD_{02} and QD_{03}. Although QD_{02} contains three PNP Darlington pairs, and QD_{03} contains three NPN Darlington pairs, only one set of driver circuits is shown in any detail. The spindle-drive signal to QD_{02} and QD_{03} is discussed in Sec. 6.13.

The timing chart in Fig. 6.5 shows switching of the three transistors to control current through the three coils of the three-phase spindle motor when playing a standard play (CAV) laserdisc. *During period A*, QD_{03a} and QD_{03b} are forward-biased; current flows from the 16-V source through QD_{02b}, LB, LA, and QD_{03a}. *During period B*, current continues to flow through LA and QD_{03a}, but the current from the 16-V source now flows through QD_{02c} and LC. *During period C*, current flows through QD_{02c}, LC, LB, and QD_{03b}. This cycle continues as the spindle motor makes one complete revolution (time periods D through I) and continues to repeat the cycle as the motor spins.

6.5 SYSTEM CONTROL

Figure 6.6 shows the relationship of the main system-control microcomputer QU_{01} and the peripheral devices, including the sub or slave microcomputers. (Microprocessors are called microcomputers in our player.) Figure 6.7 shows the interface of QU_{01} and the spindle TBC, focus and tracking servo circuits, and input-output expander QU_{03}.

As shown in Fig. 6.6, QU_{01} controls all of the player circuits through a *communications bus*. This includes the keys and display microcomputer QF_{01}, the digital-audio microcomputer Q_{304}, and the display decoder QT_{01}. QF_{01} receives commands from either the remote unit or the front-panel functions keys and sends data to the fluorescent tube display. Q_{304} controls decoding of the digital audio. QT_{01} displays data on the TV or monitor screen.

As shown in Fig. 6.7, QU_{01} communicates with the input-output expander QU_{03} through *data and address buses*. With the expansion capability of QU_{03}, QU_{01} is able to control additional player functions such as digital memory, video demodulator, analog audio, and loading motor. In addition, QU_{01} is able to re-

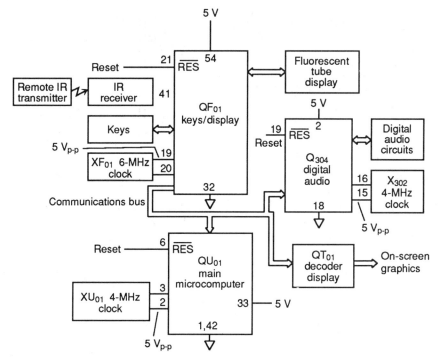

FIGURE 6.6 Relationship of main system-control microcomputer QU_{01} and peripheral devices.

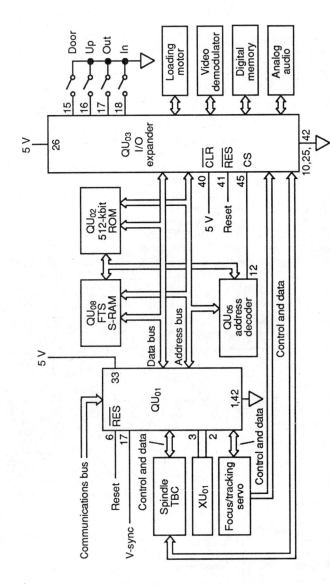

FIGURE 6.7 Interface of QU_{01} and other control circuits.

ceive data through QU_{03} from the door, up, in, and out switches and to make decisions based on these switch inputs.

The bidirectional data and address buses are also linked with two memories: a 512K ROM QU_{02} and a favorite-track selection (FTS) RAM QU_{08}. Addressing of the various devices is controlled by address controller QU_{05}.

6.5.1 Reset, Power, Ground, and Clock Functions

All of the microcomputers shown in Figs. 6.6 and 6.7 require power and ground connections, as well as clock and reset signals (as do all microprocessors). As discussed in Chap. 9, if you suspect a malfunction in system control, or with any microprocessor, one of the first troubleshooting steps is to check all of these inputs. For example, the main microcomputer QU_{01} requires 5 V at pin 33, a ground at pins 1 and 42, a 4-MHz clock at pins 2 and 3, and a reset signal at pin 6. If any of these inputs is absent or abnormal, the microcomputer will not function properly (if at all).

The power, ground, and clock functions are obvious and need not be discussed here. However, note that all microcomputers in our player use a common reset signal (called $\overline{RES}$). Figure 6.8 shows the reset circuits and related timing charts.

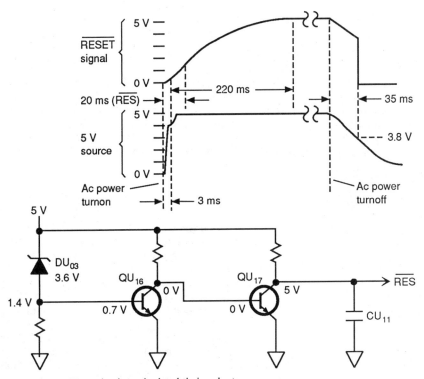

FIGURE 6.8 Reset circuits and related timing charts.

Each microcomputer requires a low (0 to 0.8 V) at the $\overline{\text{RES}}$ input for about 20 ms after power is applied. When power is first applied to the player, 5 V is applied to the power-on reset circuits shown in Fig. 6.8. Capacitor CU_{11} charges up to 5 V in about 220 ms, applying a high to the $\overline{\text{RES}}$ line. However, $\overline{\text{RES}}$ is low (0 to 0.8 V) for about 20 ms, fulfilling the reset requirements for all microcomputers in the system.

If power is removed, the 5-V source dissipates slowly as the power-supply filter capacitors discharge. To pull the $\overline{\text{RES}}$ line low (to about 3.8 V) in a short time, QU_{16} becomes reverse-biased and QU_{17} turns on, pulling the $\overline{\text{RES}}$ line low. Thus, if power is momentarily lost, or if the power supply falls too low, all of the microcomputers are reset, preventing a fault or error from occurring.

6.6 START-UP AND LASER CONTROL

Figure 6.9 shows the system-control functions involved during start-up, while Fig. 6.10 shows the related focus and laser-control circuits. Figure 6.11 shows the disc-detection sequence during start-up.

6.6.1 Mechanical Sequence

When the loading tray is pulled into the player, QU_{01} instructs the circuit to look for a disc (disc detection). As the tray is pulled in, the pick-up drive mechanism scans (using a high-speed scan) from the park position in an outward direction (for about 1 s) to a position that allows for detection of an 8- or 12-in (Laservision) disc.

Next, the loading motor turns on to place the tray in the up position. When the tray reaches the up position, the up switch opens, pin 16 of QU_{03} goes high, and the tray motor reverses direction to place the tray in the play position. This action, known as *clamping*, allows the disc to be seated properly on the spindle.

6.6.2 Focus Search

After the tray goes into the play position (down), the laser turns on and the objective lens performs a focus search operation to detect the presence of an 8- or 12-in (Laservision) disc. If focus is achieved, the presence of a Laservision disc is indicated, and play can be activated.

If focus is not found, the pickup drive moves inward until the $\overline{\text{PARK}}$ line goes high (about 5 V) and continues to move inward for 300 ms after $\overline{\text{PARK}}$ goes high. (The $\overline{\text{PARK}}$ line at pin 11 of QU_{03} is controlled by a park switch, actuated by the pickup drive mechanism.)

The laser again turns on and the objective lens again performs a focus search to detect the presence of a 3- or 5-in disc (CD or CDV Single). If focus is still not found (no disc of any size installed), the pickup drive returns to the park position, and play cannot be initiated.

TYPICAL CDV PLAYER CIRCUITS   6.13

Tray position	Door (15) PC$_4$	Up (16) PC$_5$	Out (17) PC$_6$	In (18) PC$_7$
Down	L	H	H	L
Up	L	L	H	H
Out	H	L	L	H

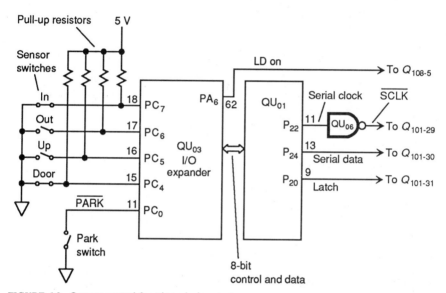

FIGURE 6.9 System-control functions during start-up.

6.6.3 Mechanical Sensor Switches

The tray-position truth table in Fig. 6.9 shows the logic of the expander QU_{03} inputs under various conditions. When a mechanical-sensor switch is closed, a low (ground) is applied to the QU_{03} input. When the switch is open, a high (5 V) is applied through the corresponding pull-up resistor.

With no disc installed, and with power applied, the sensor switches are as follows: in and door switches closed (low) and up and out switches open (high). Therefore, pins 15 and 18 are low (ground) and pins 16 and 17 are high (about 5 V). Pin 11 receives a low from the park switch.

6.6.4 Focus and Tracking Control

When a command to read a disc is executed from QU_{01}, pin 62 (LD On) of expander QU_{03} goes high to turn the laser on. Control of focus and tracking is provided by serial data (8 bit) through the serial-clock ($\overline{SCLK}$), serial-data, and latch

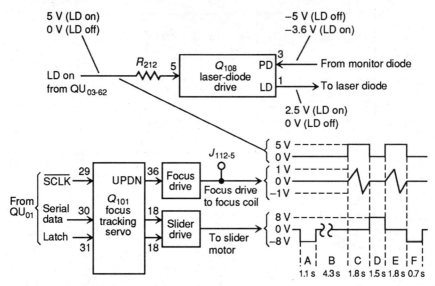

FIGURE 6.10 Focus and laser-control circuits during start-up.

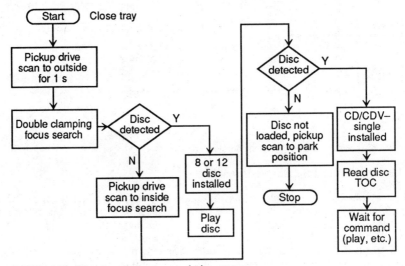

FIGURE 6.11 Disc-detection sequence during start-up.

buses. The serial-clock signal is inverted by NAND gate QU_{06} and coupled to focus-tracking servo Q_{101}.

6.6.5 Laser Control

The laser-diode on (LD On) signal is applied to laser drive IC Q_{108} through R_{212}, as shown in Fig. 6.10. When pin 5 of Q_{108} goes high, the laser is turned on

TYPICAL CDV PLAYER CIRCUITS 6.15

(through pin 1 of Q_{108}). The intensity of the laser diode is monitored by a monitor diode, which develops a voltage that is proportional to the laser intensity. The laser light output is about 0.3 mW (with a maximum of about 0.8 mW).

The monitor-diode signal applied to pin 3 of Q_{108} regulates drive to the laser. When the laser is on, pin 1 of Q_{108} (LD) is about 2.5 V, while pin 3 (PD) is typically −3.6 V. Pin 3 is −5 V when the laser is off.

6.6.6 Objective Lens Control

The serial data bits from QU_{01} are fed to Q_{101} to control focus drive and slider-drive circuits during start-up. The signal at pin 36 of Q_{101} (UPDN) controls the objective lens during start-up focus, which is initiated (1) when the loading tray is retracted into the player, (2) when power is applied to the player, or (3) when play mode is activated (if a disc is loaded into the player).

6.6.7 Disc-Detection Sequence

The disc-detection timing chart in Fig. 6.10 shows what takes place as the loading tray is retracted (with no disc present). When the close-tray command is entered, the slider motor receives about −8 V for 1.1 s (period A) to move the pickup drive outward. After 4.3 s (period B), the laser turns on and the focus drive (to the focus coil) begins a focus search pattern (objective lens up and down) for 1.8 s (period C). Since there is no disc in the player, focus is not achieved.

After this first focus search (focus start-up) a second focus search is initiated. The pickup drive is moved inward by applying 8 V to the slider motor for 1.5 s (period D). During period D (also 1.8 s), the laser turns on and a focus start-up again takes place. Since there is no disc present (in our example), the pickup drive is returned to the park position by applying a −8 V to the slider motor for 0.7 s (period F).

This sequence must take place to detect the presence of a disc before play can occur. Since no disc is detected (in our example), the player cannot be placed in the play condition.

6.7 SERVO SYSTEMS

Figure 6.12 shows the four basic servo systems used in our CDV player. Note that the four photodiodes B_1 through B_4 provide error signals to the focus, tracking, and slider servos. These same photodides are used to extract video- and audio-signal information from the disc, as described in Sec. 6.11. The tilt servo receives error signals from a separate set of diodes.

The focus error signal developed by the focus circuits (Sec. 6.8) is fed to the focus drive circuits. In turn, the focus drive circuits provide the drive to the focus coil on the pick-up drive mechanism to keep the laser beam focused on the disc.

The tracking servo (Sec. 6.9) receives tracking-error (TR) signals from the photodiodes and develops the tracking-error signal. The TR error signal is fed to

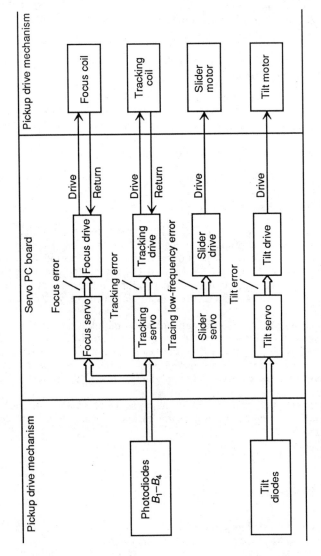

FIGURE 6.12 Four basic CDV player servo systems.

the tracking-drive circuit, which applies drive to the tracking coil on the pick-up. The coil provides a TR return to the tracking drive and to the slider servo.

The slider servo (Sec. 6.9) provides low-frequency (LF) tracking corrections. The TR LF error signal is sent to the slider-drive circuit to drive the slider motor.

The tilt servo (Sec. 6.10) uses the error signal from the tilt diodes to make tilt corrections on the pickup drive mechanism.

6.8 FOCUS SERVO

Figures 6.13 through 6.17 show the focus servo circuits. The following paragraphs describe each of the focus servo circuits.

6.8.1 Focus Error

As shown in Fig. 6.13, two signals are developed in the focus servo circuit to keep the laser beam focused on the disc. The $\overline{\text{DSUM}}$ (diode sum) and FE (focus error) signals are derived from the outputs of photodiodes B_1 through B_4, which are part of the pick-up assembly. All of the photodiode signals are summed, inverted, and doubled in op amp Q_{103b}.

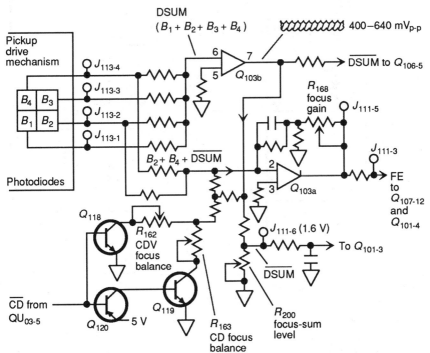

FIGURE 6.13 Focus-error circuits.

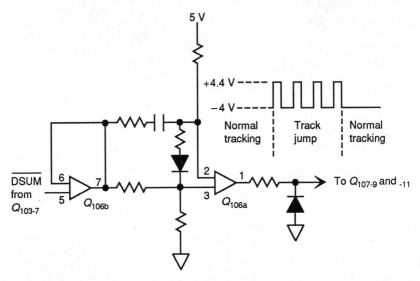

FIGURE 6.14 Focus-sum control circuits.

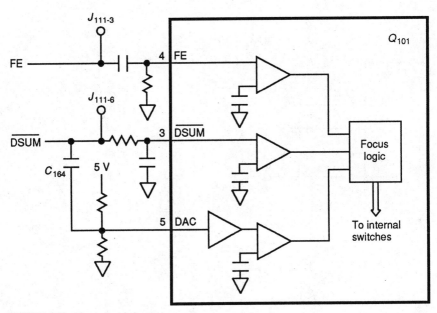

FIGURE 6.15 Focus-logic circuits.

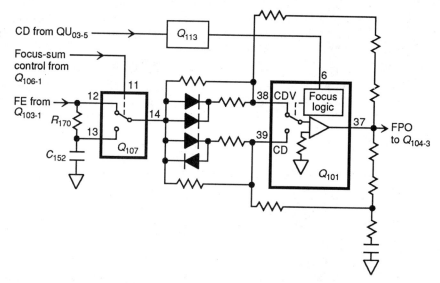

FIGURE 6.16 Focus-output circuits.

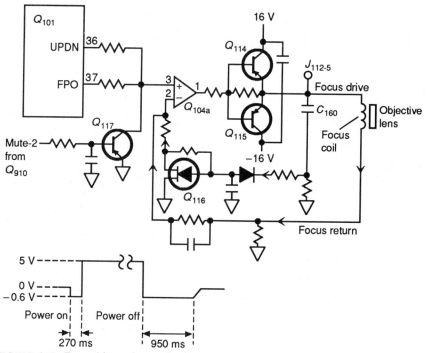

FIGURE 6.17 Focus-drive and mute circuits.

The $\overline{\text{DSUM}}$ signal is applied to pin 5 of Q_{106}, to pin 2 of Q_{103a}, and to pin 3 of Q_{101}. Note that the signal to pin 3 of Q_{101} is set by focus sum level adjustment R_{200}. In simple terms, R_{200} is adjusted (using a test disc) to provide 1.6 V at pin 6 of J_{111}. Full adjustment details are given in Chap. 9.

Two of the photodiode signals, B_2 and B_4, are applied to Q_{103a} along with the inverted $\overline{\text{DSUM}}$ signal. The resultant FE signal at the output of Q_{103} is applied to pin 12 of Q_{107} and pin 4 of Q_{101}. Note that the gain of Q_{103a} is set by focus gain control R_{168}. In simple terms, R_{168} is adjusted to provide correct focus servo loop gain (as described in Chap. 9).

6.8.2 Focus Balance

The focus balance circuits adjust the focus according to the type of disc being played (as determined by system control). As shown in Fig. 6.13, a CD signal from pin 5 of system-control expander QU_{03} is applied to switching transistors Q_{118} through Q_{120}. If a CD is played, the $\overline{\text{CD}}$ line at $QU_{03\text{-}5}$ is low. If the disc is a CDV (Laservision), $QU_{03\text{-}5}$ is high.

With the $\overline{\text{CD}}$ line high, Q_{118} is on while Q_{119} and Q_{120} are off. This places the CDV focus balance control R_{162} in the circuit. When $QU_{03\text{-}5}$ is low, Q_{119} and Q_{120} are on and Q_{118} is off, placing the CD focus balance control R_{163} in the circuit.

The focus balance controls R_{162} and R_{163} are adjusted to keep the objective lens at an optimum distance from the disc. If the proper balance is not maintained, while playing a CD or the audio portion of a CDV Single, the audio is noisy. If proper distance is not maintained while playing a CDV, there can be crosstalk in the video. The adjustment procedures for both focus balance controls R_{162} and R_{163} are given in Chap. 9.

6.8.3 Focus Sum Control Signal

As shown in Fig. 6.13, the $\overline{\text{DSUM}}$ signal is applied to pin 5 of Q_{106b}. As shown in Fig. 6.14, Q_{106b} acts together with Q_{106a} as a unity-gain op amp to develop a control signal. This control signal is applied to pins 9 and 11 of Q_{107} and serves to change the focus level during normal and special tracking.

During normal tracking, pin 1 of Q_{106a} is a steady -4 V. During track jumping (search, fast forward, fast reverse, track loss, etc.), the signal pulses, or toggles, between -4 and $+4.4$ V. This is discussed further in Sec. 6.8.5.

6.8.4 Focus Logic

As shown in Fig. 6.15, the FE and $\overline{\text{DSUM}}$ signals are applied to the focus logic circuit, which is part of the focus tracking slider (FTS) servo IC Q_{101}. The focus logic circuit turns the servo loop on when certain conditions (which indicate focus lock) are met during start-up (Sec. 6.6).

The focus lock condition is met when the $\overline{\text{DSUM}}$ input (pin 3) to Q_{101} reaches 0.4 V and when the FE input (pin 4) reaches 0.3 V. If the $\overline{\text{DSUM}}$ signal falls below 0.4 V because of a focus loss, Q_{101} instructs the focus servo to open, and a focus search pattern is activated. (A damaged disc or excessive vibration are the most common causes for a loss of focus.)

Note that the pin 5 of Q_{101} is derived from the $\overline{\text{DSUM}}$ signal (through C_{164}) and is applied to a comparator within Q_{101}. This signal provides for main-beam on-off track detection. If the signal at $Q_{101\text{-}5}$ is below a threshold of 0.5 V, this indicates that the main beam is off track.

6.8.5 Focus Output

As shown in Fig. 6.16, the focus sum control signal (Sec. 6.8.3) is applied to pin 11 of Q_{107} to control the input sent to the CD or CDV focus output circuit. When the focus sum signal is low, pin 12 of Q_{107} is connected to pin 14, and the FE signal is passed directly. When the focus sum signal is high, pin 13 is connected to pin 14, and the FE signal is attenuated by R_{170} and C_{152}. Since the focus sum control signal toggles between high and low during track jumping, FE is also attenuated during track jumps but remains constant during normal tracking.

The FE signal is applied to the FTS servo Q_{101} through feedback phase-compensation networks at pins 38 and 39. The correct network is selected according to the type of disc being played (as determined by system control). A CD signal from pin 5 of system-control expander QU_{03} is applied to pin 6 of Q_{101} through Q_{113}.

If a CD is played, the $\overline{\text{CD}}$ line at $QU_{03\text{-}5}$ is low, connecting the network at pin 39. If a CDV is played, $QU_{03\text{-}5}$ is high, and the network at pin 38 is used. Either way, the FE signal is applied to the focus drive and mute circuit through pin 37 of Q_{101} as a focus output (FPO) signal.

6.8.6 Focus Drive and Mute

As shown in Fig. 6.17, both the FPO and UPDN (Sec. 6.6.6) signals are applied to the focus coil of the objective lens through Q_{104a}, Q_{114}, and Q_{115}. Note that both of the signals are muted during power-on and power-off by a mute-2 signal (Sec. 6.3), which is applied through Q_{117}. This prevents the objective lens from moving up and down excessively during power on or off.

Feedback is provided to the inverting input of Q_{104a} (from a focus return loop) to stabilize focus servo operation. The FPO signal at the focus coil is also returned to a high-frequency focus limiter loop through C_{160}. This lowers the loop gain when the focus error signals are high frequency (to further stabilize the focus servo operation).

6.9 TRACKING AND SLIDER SERVO

Figures 6.18 through 6.21 show the tracking-servo circuits, while Fig. 6.22 shows the slider servo. The following paragraphs describe the functions of these two interrelated servos.

6.9.1 Tracking Error

As in the case of a CD (Figs. 2.8 through 2.10), two photodiodes are used to read the secondary (or radial) beams reflected from a CDV disc to provide radial

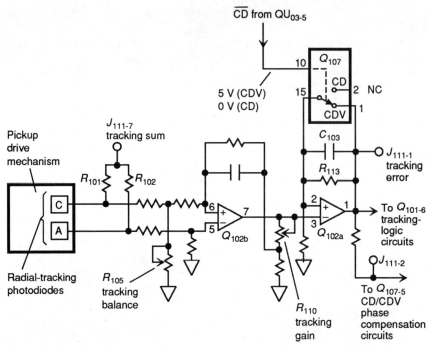

FIGURE 6.18 Tracking-error circuits.

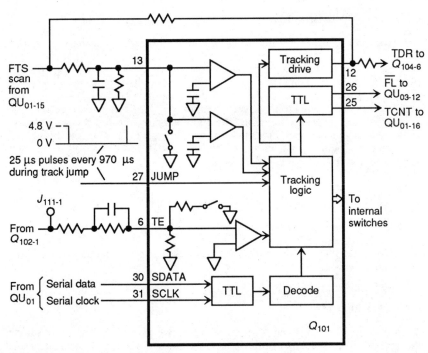

FIGURE 6.19 Tracking-logic circuits.

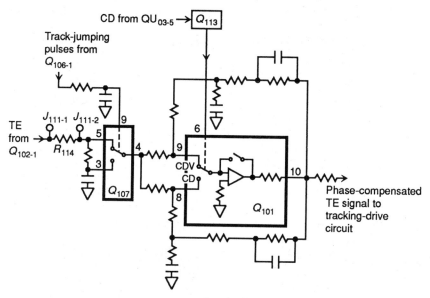

FIGURE 6.20 CD and CDV phase-compensation circuits.

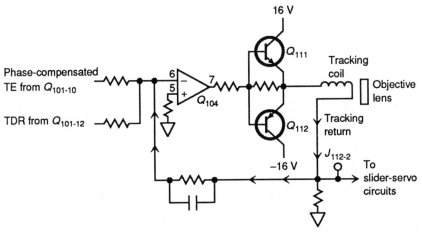

FIGURE 6.21 Tracking-drive circuits.

tracking. As shown in Fig. 6.18, the two radial-tracking photodiodes are labeled A and C (whereas the four main photodiodes are labeled B_1 through B_4). The signals from photodiodes A and C are sent to a difference amplifier in Q_{102b} to produce a tracking-error signal at pin 7. The signals are also added through R_{101} and R_{102} at test point J_{111-7} to produce a tracking-sum signal.

The tracking-error signal is applied to the noninverting input (pin 3) of tracking and equalization (TR/EQ) amplifier Q_{102a}. The feedback equalization (or gain) for

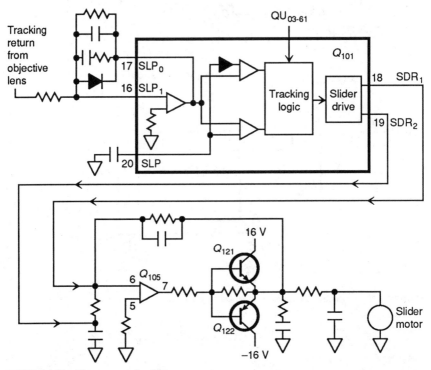

FIGURE 6.22 Slider-servo circuits.

Q_{102a} is set by the type of disc being played (as discussed in Sec. 6.8.2). If a CD or the audio portion of a CDV (Gold CD) is being played, the $\overline{CD}$ line is low, placing the switch within Q_{107a} in the CD condition, with C_{103} and R_{113} in the circuit. When a CDV (or the video portion of a Gold CD) is being played, the $\overline{CD}$ line is high, placing the Q_{107} switch in the CDV position (with C_{103} and R_{113} out of the circuit, shorted by the Q_{107} switch).

The tracking-error signal can be checked at J_{111-1}. Tracking-balance control R_{105} is used to adjust the signal at J_{111-1} so that the signal is zero when tracking is properly balanced, as described in Chap. 9. Tracking-gain control R_{110} is used to set the tracking-servo loop gain, also described in Chap. 9. Note that the tracking-error signal is applied to both the tracking-logic circuit (Sec. 6.9.2) and the CD/CDV phase-compensation circuit (Sec. 6.9.3) before the error signal is applied to the objective lens (Sec. 6.9.4).

6.9.2 Tracking Logic

As shown in Fig. 6.19, the tracking-logic circuit monitors several signals that are used to control the tracking servo. The tracking-logic circuit also receives commands (forward, reverse, start, stop, etc.) from system control through the serial data-bus lines (Sec. 6-5).

Pin 6 of FTS servo Q_{101} receives the tracking-error (TE) signal to determine tracking conditions. Q_{101} monitors the frequency of the TE signal to open or close the tracking-servo loop or to change the servo-loop characteristics. When operations such as search, fast forward, or fast reverse are activated, pin 27 of Q_{101} receives pulses from system-control expander QU_{03} to move the pickup-drive mechanism in the proper direction. During such an operation, the tracking-servo loop must be open to allow jumping or skipping over tracks on the disc.

The tracking-logic circuit also provides tracking information to system control. The tracking-logic circuit provides a tracking count (TCNT) signal to pin 16 of system-control microprocessor QU_{01}, providing QU_{01} with information to determine the position of the laser beam on the disc.

The focus-lock ($\overline{FL}$) signal is applied to QU_{03} at pin 12, indicating the condition (locked or unlocked) of the focus-servo loop. The $\overline{FL}$ signal is low (0 V) when focus is achieved (locked) and high (5 V) when the beam is not in focus (unlocked). Note that the tracking circuit is not activated until the focus servo loop is locked.

QU_{01} provides an FTS scan signal at pin 12 of Q_{101} to open the tracking-logic circuit during slow and medium scan operation. When pin 13 of Q_{101} is at 0.2 (ST), the tracking loop opens until the frequency of the TE signal at pin 6 drops below a certain frequency (determined by the scan speed). While the tracking loop is open, a brake pulse is taken from the tracking-drive (TDR) output at pin 12. The TDR signal is applied to the tracking-drive circuits (Sec. 6.9.4).

6.9.3 CD and CDV Phase Compensation

As shown in Fig. 6.20, the TE signal is applied to pin 5 of Q_{107a}. The control signal at pin 9 is the same signal used in the focus-servo circuit, and it operates the Q_{107} switch as described in Secs. 6.8.3 and 6.8.5. That is, the TE signal is attenuated during track jumping but remains steady during normal tracking.

In either condition, the TE signal is applied to Q_{101} through feedback phase-compensation networks at pins 8 and 9. The correct network is selected according to the type of disc being played (as determined by system control). A CD signal from pin 5 of QU_{03} is applied to pin 6 of Q_{101} through Q_{113}. If a CD is played, the CD line at QU_{03-5} is low, connecting the network at Q_{101-8}. If a CDV is played, QU_{03-5} is high, and the network at Q_{101-9} is used. Either way, the TE signal is applied to the tracking-drive and slider circuits through pin 10 of Q_{101}.

6.9.4 Tracking Drive

As shown in Fig. 6.21, both the tracking-drive (TDR) and tracking-error (TE) signals (Secs. 6.9.2 and 6.9.3) are applied to the tracking coil of objective lens through Q_{104}, Q_{111}, and Q_{112}. Feedback is provided to the inverting input of Q_{104} (from a tracking-return loop). This feedback signal is also applied to the slider-servo circuits.

6.9.5 Slider Servo

As shown in Fig. 6.22, the tracking-return signal (used as a stabilizing feedback control, as described in Sec. 6.9.4) is also applied to the slider motor through

6.26 CHAPTER SIX

Q_{101}, Q_{105}, Q_{121}, and Q_{122}. As the laser beam moves over the disc, the tracking servo applies current to the objective lens in an attempt to keep the beam on track. Since there is a limit to how far the objective lens can move (right or left), the slider servo is used to move the entire pickup drive mechanism (including the objective lens) across the disc (toward the outer tracks during play).

When a track-jumping operation (search, fast forward, fast reverse, track loss, etc.) is necessary, the tracking loop is open, and the tracking-logic circuit controls movement of the pickup mechanism, as described in Sec. 6.9.2.

6.10 TILT SERVO

Figure 6.23 shows the tilt-servo circuits. These circuits ensure that the laser beam tracks perpendicularly to the disc plane. This is necessary for the laser beam to

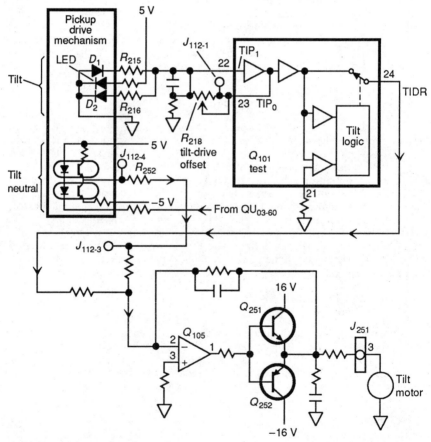

FIGURE 6.23 Tilt-servo circuits.

be reflected back into the objective lens and be picked up by the signal photodiodes (B_1 through B_4, Fig. 6.13).

Note that the tilt servo is active only when 8- and 12-in discs are played since large discs are more likely to be warped. When small discs (3- or 5-in) are played, the pickup drive remains in a neutral position as sensed by the tilt-neutral opto-transistors located on the drive mechanism.

An LED emits a beam that is reflected by the disc and picked up for the tilt photodiodes D_1 and D_2. Each diode produces voltage in proportion to the intensity of the light received. If D_1 and D_2 receive an equal amount of light (disc flat), no current flows through R_{215} and R_{216} since the polarity of the diodes is opposing. If the disc is not flat, the diodes receive unequal light and a difference-signal is generated.

The difference signal is applied to pin 22 of Q_{101}, which develops a tilt-drive (TIDR) signal at pin 24. The TIDR signal is applied to the tilt motor through Q_{105}, Q_{251}, and Q_{252} as necessary to offset any warping of the disc. The tilt-drive signal-offset is adjusted by R_{218} (Chap. 9). R_{218} is adjusted for zero offset when there is no tilt.

If a 3- or 5-in disc is played, pin 60 of QU_{03} goes low, turning on a pair of tilt-neutral opto-transistors. Under these conditions, -5 V is applied to pin 2 of Q_{105} through the tilt-neutral opto-transistors and R_{252}. This disables the tilt-drive circuit.

6.11 DISC SIGNAL PROCESSING

Figures 6.24 through 6.26 show the disc signal-processing circuits. The following paragraphs describe each of the signal-processing functions.

6.11.1 RF Preamp

As shown in Fig. 6.24, disc video and audio information picked up by the four photodiodes B_1 through B_4 is amplified by an RF preamp. Both the photodiodes

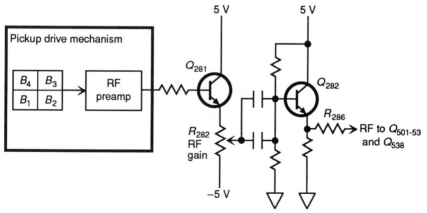

FIGURE 6.24 RF preamp circuits.

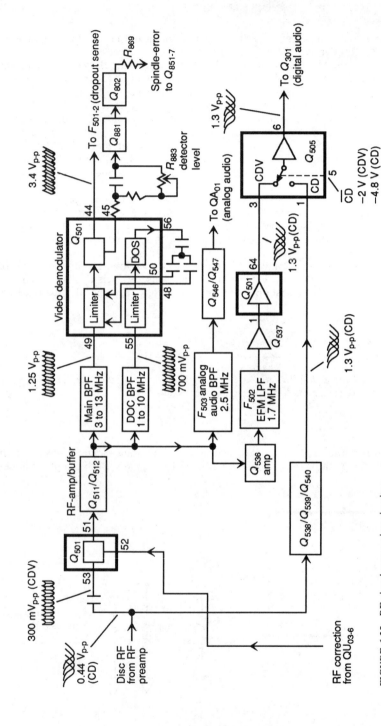

FIGURE 6.25 RF signal-processing circuits.

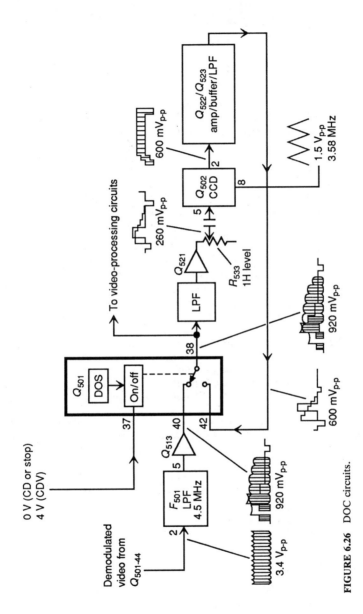

FIGURE 6.26 DOC circuits.

and preamp are part of the pickup mechanism. The amplified disc signal is applied to the RF signal-processing circuits through Q_{281} and Q_{282}. The signal gain is set by RF-gain control R_{282}.

6.11.2 RF Signal Processing

As shown in Fig. 6.25, the RF (or EFM) signal is applied to both the video and audio circuits after amplification by the preamp.

The main disc signal passes through an RF-correction circuit in Q_{501}, amplifier Q_{511}, buffer Q_{512}, a main BPF (that passes signals between 3 and 13.5 MHz), a limiter in Q_{501}, and the video demodulator in Q_{501}. The demodulated video is passed through a DOC circuit (Fig. 6.26) before the video is applied to the video-processing circuits (Sec. 6.12).

The disc signal also passes through a DOC BPF (that passes RF signals between 1 and 10 MHz), a limiter in Q_{501}, and a drop-out sense (DOS) circuit in Q_{501}. The resultant signal is added to the main disc signal in Q_{501}.

The analog-audio portion of the RF signal (2.5 MHz) is selected by BPF F_{503} and applied to the analog audio circuits (Sec. 6.16) through amplifier Q_{546} and buffer Q_{547}.

The EFM portion of the RF signal (below 1.7 MHz) is extracted by low-pass filter (LPF) F_{502} after amplification by Q_{536}. The EFM signal is then applied to the CD and CDV switch in Q_{505} after being buffered by Q_{537} and Q_{501}.

Q_{505} selects the EFM input to be processed by the digital audio circuits (Sec. 6.17). If a CD is played, the CD line goes low (-4.8 V) to place the Q_{505} switch in the CD position. When a CD, or the audio portion of a CDV Single, is played, the EFM from the disc is routed directly from the RF preamp through amplifier and buffers Q_{538} and Q_{540} to the digital audio circuit (without processing).

If a CDV (which contains digital audio) is played, the CD line is high (-2 V), placing the Q_{505} switch in the CDV position. Under these conditions, EFM from the disc is routed to the digital audio circuit through the correction and processing functions shown in Fig. 6.25.

The main demodulated video is also used to provide a spindle-error signal that is applied to the spindle-drive and feedback circuits ($Q_{851\text{-}7}$, Fig. 6.36) through R_{883}, Q_{881}, Q_{802}, and R_{869}. The spindle-drive and feedback system reduces the time-base error in reproduced video by detecting the reproduced composite-sync and then accelerating or decelerating (braking) the spindle motor as necessary to reduce the time-base error. Detector level R_{883} sets the level of the spindle-error signal.

6.11.3 DOC Circuit

As shown in Fig. 6.26, the main video signal is applied through a DOC circuit before the signal is applied to the video-processing circuits (Sec. 6.12). The demodulated video at pin 44 of Q_{501} is applied to the DOC circuits through 4.5-MHz LPF F_{501}, Q_{513}, and a switch in Q_{501}.

The Q_{501} switch normally connects pins 38 and 40 of Q_{501} to pass the video signal to the video-processing circuits. If a video dropout occurs (say because of dirt or a scratch on the disc) the DOS circuits in Q_{501} sense the loss of video signals and move the Q_{501} switch so that pins 38 and 40 are connected. This applies the previous horizontal line (1H) of video to the video-processing circuits, thus making up for the loss of video.

The horizontal line of video is developed by the LPF at pin 38 of Q_{501} and at Q_{521}, Q_{502}, Q_{522}, and Q_{523}. The LPF removes chroma from the video signal and produces a luma signal which is amplified by Q_{521}. The luma signal is then returned to pin 42 of Q_{501} through R_{533}, 1H-delay Q_{502}, and amplifier, buffer, and limiter Q_{522} and Q_{523}. Q_{502} uses charge-coupled device (CCD) techniques to delay the video by 1H. Q_{502} is synchronized to the video signal by a 3.58-MHz clock at pin 8. R_{533} sets the level of the 1H signal (Chap. 9).

Note that the DOC circuit is inhibited by a low at pin 37 of Q_{501} when a CD is being played or during stop. When a CDV is played, pin 37 is high (4 V), permitting the Q_{501} DOS circuit to control the Q_{501} DOC switch.

6.12 VIDEO PROCESSING

Figures 6.27 through 6.32 show the video-processing circuits. The following paragraphs describe each of the signal-processing functions.

6.12.1 Time-Base Correction

As shown in Fig. 6.27, video from pin 38 of Q_{501} (Fig. 6.26) is applied to TBC circuits before being applied to the spindle-servo (Sec. 6.13) and video-distribution circuits (Sec. 6.14). The TBC circuits use CCD techniques for time-base correction. A TBC error loop (composed of sync and data separation, time-base error detection, and VCO blocks) develops a TBC error signal by comparing the playback-horizontal (PB-H) signal to a horizontal-reference (H-ref) signal.

The time-axis correction block is a coarse-correction circuit that removes horizontal jitter from the playback video. To provide fine correction of the chroma phase, the CCD video is fed to a phase-correction block. The phase-error detection block develops a video phase-shift error (VPS ERR) signal by comparing the 3.58-MHz playback burst (detected from the CCD video) to a 3.58-MHz reference signal. The VPS ERR signal is applied to the phase-correction block, along with the CCD video to provide correction of chroma-phase jitter.

6.12.2 Time-Axis Correction

As shown in Fig. 6.28, the time-axis correction circuit receives video from pin 38 of Q_{501}. The video is applied to Q_{504} through Q_{514} and Q_{515} and C_{536} and C_{537}. The TBC error signal (Sec. 6.12.5) is applied to VCO Q_{503} to develop a variable clock signal (9 to 14 MHz), which is applied to Q_{504} to control timing of the clock driver.

The clock driver controls timing of the CCDs to provide a variable delay of the video signal. The VCO center adjust R_{549} is used to provide an optimum average (or center) delay from the input video (at pin 9 of Q_{504}) to the CCD video (at the emitter of Q_{527}). Two CCDs are used in parallel to get sufficient bandwidth and delay. The average delay is adjusted to 70.7 μs (1H plus 7.2 μs) ± 0.1 μs, as described in Chap. 9. Note that a color-lock failure (or a slow color lock) after a search operation may indicate the need for adjustment of R_{549}.

The time-base corrected video (CCD video) at pin 16 of Q_{504} is applied to the

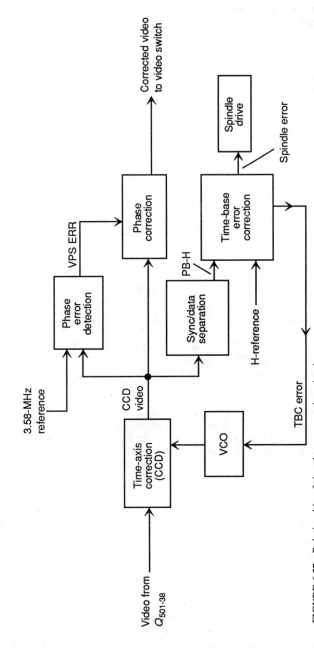

FIGURE 6.27 Relationship of time-base correction circuits.

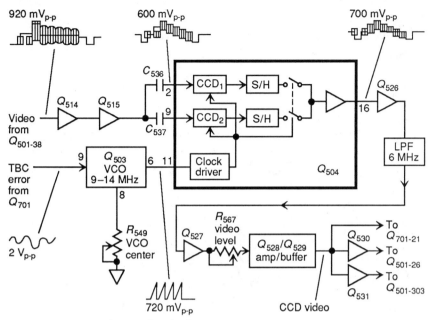

FIGURE 6.28 Time-axis correction circuits.

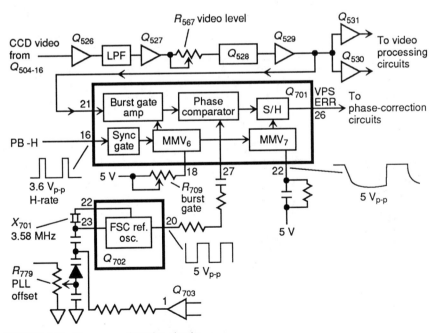

FIGURE 6.29 Phase-error detection circuits.

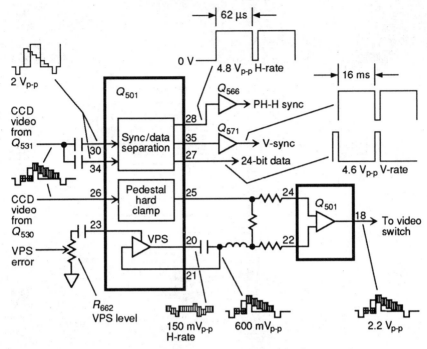

FIGURE 6.30 Phase-correction and sync and data-separation circuits.

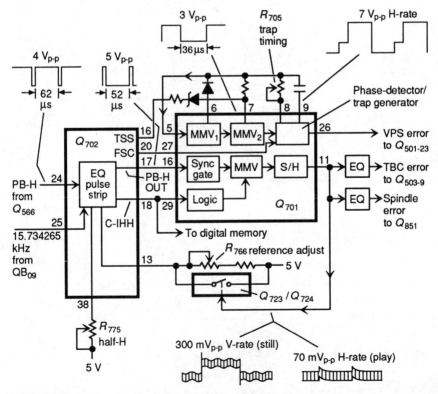

FIGURE 6.31 TBC-error and spindle-error detection circuits.

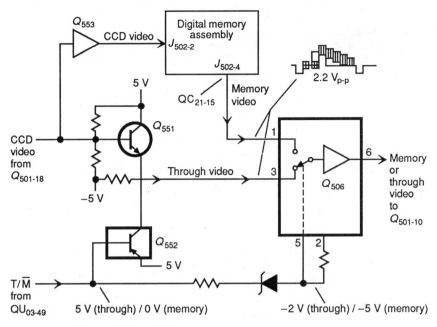

FIGURE 6.32 Video-switch circuits.

phase-error detection circuits (Fig. 6.29) and the sync and data-separator and phase-correction circuits (Fig. 6.30) through Q_{526}, a 6-MHz LPF, Q_{527}, R_{567}, and Q_{528} through Q_{531}. Video-level adjustment R_{567} is adjusted to provide a video level (pedestal to 100 percent white) of 0.71 $V_{p\text{-}p}$ at the video-output jack (with 75-Ω termination), as described in Chap. 9.

6.12.3 Phase-Error Detection

As shown in Fig. 6.29, CCD video is applied to the burst-gate amplifier of Q_{701} at pin 21. The burst-gate amplifier supplies the burst extracted from the CCD video to the Q_{701} phase comparator where the burst is compared with a 3.58-MHz reference (FSC) from Q_{702}.

The phase-comparator output is applied to the phase-correction circuits (Fig. 6.30) through a sample and hold (S/H) circuit in Q_{701}, as the VPS error signal at pin 26 of Q_{701}. The S/H circuit is clocked by MMV_7. The burst-gate adjust R_{709} is adjusted to provide a 1-μs delay from the leading edge of the video signal (at the emitter of Q_{527}).

6.12.4 Phase Correction and Sync and Data Separation

As shown in Fig. 6.30, CCD video is applied to a phase-correction circuit in Q_{501} through a pedestal hard-clamp circuit in Q_{501} (pins 24, 25, 26). The VPS error signal is also applied to the Q_{501} phase-correction circuit through VPS level adjustment R_{662} and an amplifier in Q_{501}. R_{662} is adjusted to provide proper color phase, as described in Chap. 9. The output of the phase-correction circuit is applied to the video switch (Fig. 6.32).

The sync and data-separation circuit of Q_{501} separates PB-H sync (pin 28), vertical sync (pin 35), and 24-bit data (pin 27) from the luma signal at pins 30 and 34 (from the phase-error detector, Fig. 6.29).

6.12.5 TBC-Error and Spindle-Error Detection

As shown in Fig. 6.31, both TBC-error and spindle-error signals are developed by the TBC circuit. The PB-H signal from Q_{566} (Fig. 6.30) is applied to an equalization-pulse strip circuit in Q_{702} to remove equalization pulses from the PB-H signal. The half-H adjust R_{775} is adjusted to provide 52 µs between positive-going pulses of the signal at pin 17 of Q_{702} (Chap. 9).

After the equalization pulses are removed from the PB-H signal, Q_{702} applies PB-H out and clock inhibit (C-INH) signals to Q_{701} (pins 16 and 29). The reference shift adjust R_{766} sets the width of the C-INH pulse, which is then applied to the digital memory circuit (Sec. 6.15) and to logic within Q_{701}.

A trapezoid signal is generated in Q_{701} as a reference to detect time-base errors. The signal at pin 11 of Q_{701} is developed by determining where the PB-H falls in respect to the slope of the trapezoid in the S/H circuit.

The horizontal reference signal from pin 16 of Q_{702} is adjusted by trap-timing adjustment R_{705} to get 36 µs between positive-going pulses (Chap. 9). The signal at pin 11 of Q_{701} is applied through equalization circuits to both the time-axis correction circuits of Fig. 6.28 (as the TBC error signal) and the spindle servo described in Sec. 6.13 (as the spindle-error signal).

6.12.6 Video Switch

As shown in Fig. 6.32, CCD video from pin 18 of Q_{501} (Fig. 6.30) is applied to a switch in Q_{506} through Q_{553} and the digital-memory circuits (Sec. 6.15) or directly through Q_{551}, depending on which mode is selected.

If memory video is selected (so that special effects can be used), the T/$\overline{\text{M}}$ line (from system control QU_{03-49}) is low (0 V). This biases Q_{552} on and applies 5 V to the emitter of Q_{551}, cutting Q_{551} off. At the same time, the voltage at pin 5 of Q_{506} goes low, connecting the switch to pin 1 of Q_{506}. The CCD video is thus routed through the memory circuits (for special effects) to the video-distribution circuits (Sec. 6.14).

If the memory circuits are not selected, the T/$\overline{\text{M}}$ line is high (5 V), Q_{552} is cut off, Q_{551} is on, the Q_{506} switch moves to pin 3, and CCD video (without special effects) is routed directly to the video-distribution circuits through Q_{551} and Q_{506}.

6.13 SPINDLE SERVO

Figures 6.33 through 6.36 show the spindle-servo circuits. The following paragraphs describe the functions of this servo system (used to drive the spindle motor, Sec. 6.4).

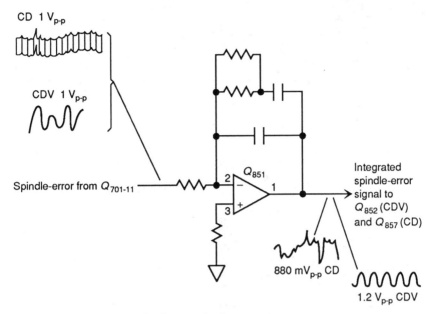

FIGURE 6.33 Spindle-drive development circuits.

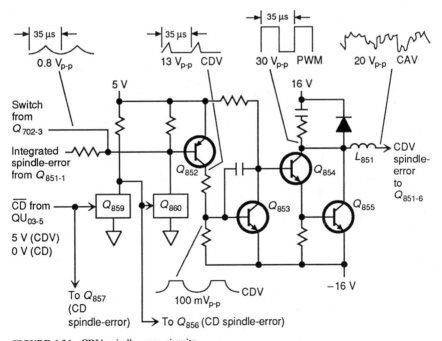

FIGURE 6.34 CDV spindle-error circuits.

6.37

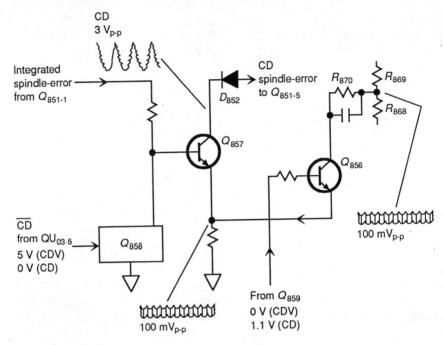

FIGURE 6.35 CD spindle-error circuits.

6.13.1 Spindle-Drive Development

As shown in Fig. 6.33, the spindle-error signal from the TBC circuit (Sec. 6.12.5) varies according to the type of disc being played (CD or CDV). The waveforms shown in Fig. 6.33 are for an audio CD and for a standard-play (CAV) laserdisc. In either case, the spindle-error signal (combined with feedback, Sec. 6.13.4) is applied to pin 2 of integration amplifier Q_{851}. The integrated output at pin 1 of Q_{851} is applied to both CDV (Sec. 6.13.2) and CD (Sec. 6.13.3) spindle-error circuits.

6.13.2 CDV Spindle-Error

As shown in Fig. 6.34, the integrated output from Q_{851} is combined with a switch signal from the TBC circuit and applied to the spindle-drive and feedback circuits (Sec. 6.13.4) through Q_{852} through Q_{855}. If CDV is selected, the $\overline{CD}$ line (from system control $QU_{03\text{-}5}$) is high (5 V). This biases Q_{859} on and Q_{860} off. The spindle-error signal then passes through Q_{852} through Q_{855} and L_{851} to the drive and feedback circuits. When a CD is played, the $\overline{CD}$ line is low, biasing Q_{859} off and Q_{860} on. (Actually, Q_{860} does not saturate but does conduct enough current to prevent the spindle signal from passing to Q_{852}.)

Both the spindle-error signal and the switch signal from the TBC circuits are applied to the base of Q_{852}. The collector signal (about 13 $V_{p\text{-}p}$) of Q_{852} is used to drive Q_{853} through Q_{855}, producing a pulse width modulation (PWM) signal of

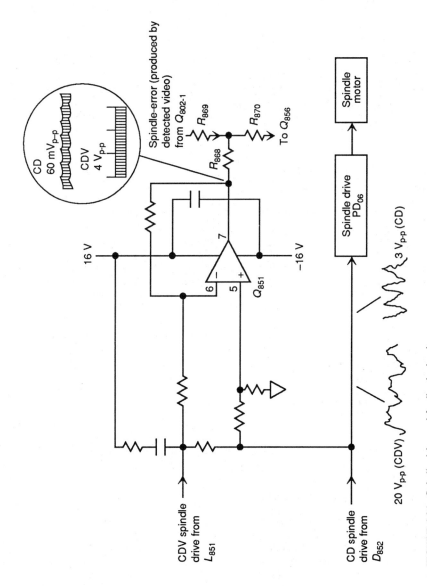

FIGURE 6.36 Spindle drive and feedback circuits.

about 30 V_{p-p} at the collectors of Q_{854} through Q_{855}. The PWM signal (representing the spindle error or drive) is applied to the drive and feedback circuits through L_{851}.

6.13.3 CD Spindle Error

As shown in Fig. 6.35, the integrated output from Q_{851} is applied to the drive and feedback circuits through Q_{857} and Q_{856}. If CD is selected, the $\overline{CD}$ line (from system control QU_{03-5}) is low (0 V). This biases Q_{858} off, allowing the spindle-error signal to be applied at the base of Q_{857}. The spindle-error signal is amplified to about 3 V_{p-p} by Q_{857} and coupled to the drive and feedback amplifier through D_{852}. Q_{856} is also biased on (by about 1.1 V from Q_{859}, Fig. 6.34), allowing the 100-mV signal at the emitter of Q_{856} to pass and be fed back.

6.13.4 Spindle Drive and Feedback

As shown in Fig. 6.36, both the CD and CDV spindle-drive signals are applied to the spindle motor (through spindle drive PD_{06}) and to the spindle-error circuits (through feedback amplifier Q_{851}). The feedback signals are combined with the spindle-error signal from Q_{701} (Fig. 6.31) and are applied to the spindle-drive development input at Q_{851} (Fig. 6.33). As discussed in Sec. 6.12.5, the spindle-error signals pass through equalization circuits before reaching Q_{851}. The spindle-error signal (produced by detecting the video composite-sync signal, Sec. 6.11.2) is applied to the feedback circuits through R_{869}.

The spindle motor has three phases and has Hall-effect sensors. The motor is controlled by the usual Hall feedback circuits located on spindle-drive board PD_{06}, as described in Sec. 6.4.

6.14 VIDEO DISTRIBUTION, NR, AND OSD

Figure 6.37 shows the noise reduction (NR) and on-screen display (OSD) circuits for our CDV player, while Fig. 6.38 shows the video distribution circuits.

6.14.1 Video Signal Processing

As shown in Fig. 6.37, the video signal at pin 6 of Q_{506} (Fig. 6.32) is applied to the video distribution circuits (Sec. 6.14.3) through video signal processor Q_{501}.

The video signal is subtracted from the low- and high-frequency signals at pins 15 and 16 (obtained from internal and external limiters controlled by the CAV/$\overline{CLV}$ line at pin 17 of Q_{501}). The resultant signal is applied to a sync-tip clamp in Q_{501}. The clamped video is applied to a video squelch (VSQ) switch, which is controlled by the VSQ signal at pin 9.

When a video disc is playing, the VSQ signal is high (5 V) and the video signal from the clamp is connected to the display-control circuits. When the player is in the stop mode, the VSQ line is low (0 V), connecting the video to a blue-

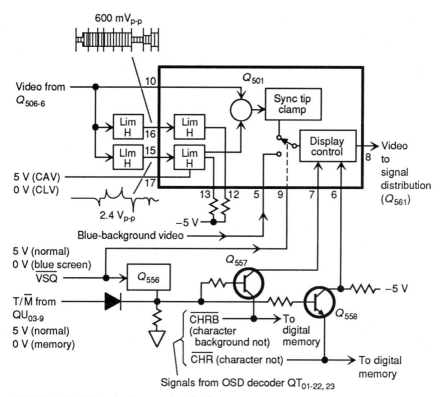

FIGURE 6.37 Video signal-processing circuits.

background video signal input at pin 5 (causing the TV or monitor screen to go blue until the player starts).

6.14.2 On-Screen Display

The OSD circuit is controlled by the display-control circuits in Q_{501} (Fig. 6.37) and by the T/$\overline{\text{M}}$ control signal from $QU_{03\text{-}9}$ (Sec. 6.12.6). Signals from the OSD decoder QT_{01} are applied to the display-control circuits through Q_{557} and Q_{558}. In turn, Q_{557} and Q_{558} are controlled by the T/$\overline{\text{M}}$ line.

When the player is not in the memory mode (no special effects), the T/$\overline{\text{M}}$ line is high. This biases Q_{557} and Q_{558} on and passes the OSD signals to the video distribution through the Q_{501} display-control circuits. When the player is in memory, the T/$\overline{\text{M}}$ line is low. This biases Q_{557} and Q_{558} off, preventing the OSD signals from passing to Q_{501}. Instead, the OSD signals are applied to the digital memory circuits (Sec. 6.15).

6.14.3 Video Distribution

As shown in Fig. 6.38, the processed video from pin 8 of Q_{501} is distributed to the RF modulator K_{501}, to the Video Out jack J_{511}, and the S-Output connector

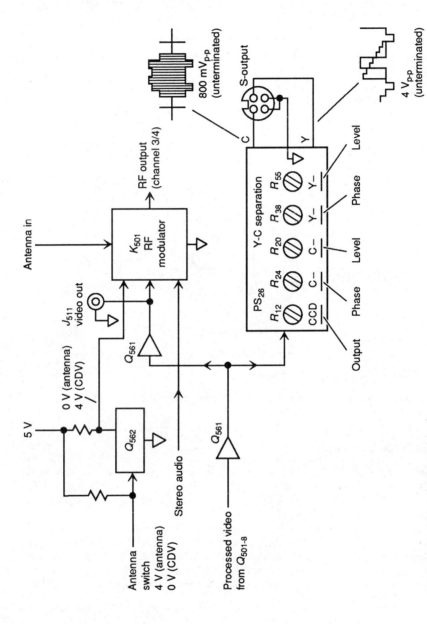

FIGURE 6.38 Video distribution circuits.

through Q_{561} and Q_{562}. The output from J_{511} is a standard composite video signal to be viewed on a monitor TV, while the signal from the S-Output connector requires a monitor capable of displaying separated Y (luma) and C (chroma) video (an S-VHS monitor). RF modulator K_{501} combines the standard composite video signal with right- and left-channel stereo audio for conversion to RF at TV channels 3 and 4, in the usual manner. K_{501} also includes an antenna input to feed external RF (from an antenna or cable TV) to the TV set.

6.15 DIGITAL MEMORY

Figure 6.39 shows the digital-memory circuits used in our CDV player. These circuits provide for the special effects described in Chaps. 1 and 3. As discussed in Sec. 6.12.6, the memory circuits can be bypassed if desired. Although the circuits are shown in block form, all of the input-output test points and signal paths are given by pin number.

CCD video is applied to the digital-memory circuits at analog-digital (A/D) converter QB_{08} through an LPF. The CCD video is converted to 8-bit data (4-bit upper and 4-bit lower) by QB_{08} at a rate of 576 samples per horizontal line. The A/D clock input at $QB_{08\text{-}12}$ is developed by memory control QB_{09}. The A/D clock is at a frequency of 9.06 MHz (which is 576 times the horizontal rate, or 576fH).

Memory control QB_{09} controls both the writing and reading operations to and from the memory IC Q_{810} (a 1-Mbit DRAM). The 18.12-MHz (1152fH) input clock at pin 10 of QB_{09} is required to write each sample into memory through the 4-bit parallel-serial bus.

Writing into memory is enabled by the write-enable (WE) inputs from system control and the WE input from pin 36 of QB_{09}. The 15-bit address bus A_0 through A_{14} is used to write the 4-bit-by-8-bit (four samples, or 32 bits) data to a four-sample RAM address in QB_{10}.

QB_{09} controls the reading of data from QB_{10} for special effects. The 8-bit memory video is converted by digital-to-analog (DAC) converter QC_{29} to an analog output at pin 6 of QC_{29}. Horizontal sync is added to the analog signal by QC_{21}, and the composite output is applied to QC_{14} where the OSD is inserted (Sec. 6.14.2).

The output from pin 1 of QC_{14} (including any OSD characters) is applied to a switch within QC_{21} through a 140-ns phase-shift circuit. The phase shift provides the correct burst-phase for the composite video signal and is operated by a switching pulse (CINV) from pin 60 of QB_{09}. The CINV pulse is developed by an edge-detect circuit and QB_{09}. The edge-detect circuit compares a 3.58-MHz reference with the burst signal of the composite video. The CINV pulse switches at the frame rate to provide a 140-ns phase-shift burst every other frame.

QC_{21} is operated by the same T/M control line used in the video switch and video distribution circuits (Figs. 6.32 and 6.37). When the line is low, the memory circuit output is applied to video distribution, and any special effects selected by system control appear in the video at pin 8 of Q_{501} (Fig. 6.37). The memory circuits are completely bypassed when the T/M line is high, thus inhibiting any special effects.

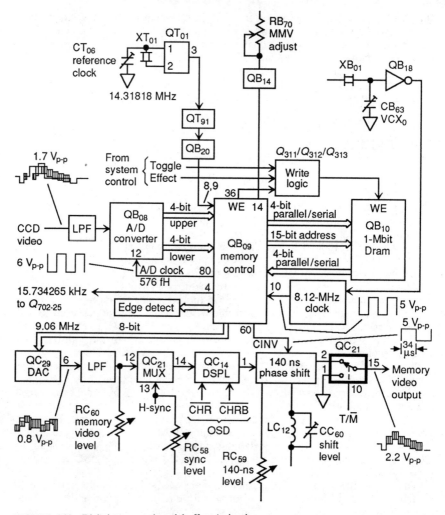

FIGURE 6.39 Digital-memory (special effects) circuits.

6.16 ANALOG AUDIO

Figure 6.40 shows the analog audio circuits for our CDV player. Again, the circuits are shown in block form, but all significant input-output test points and signal paths are given by pin number.

The audio RF or EFM signal (Fig. 6.25) is applied to the audio demodulation IC QA_{01} through two BPFs which extract FM left- and right-channel audio from the RF signal. The demodulators for the two channels are identical, so only the right channel is shown.

The audio FM signal is applied through the QA_{01} limiter to the FM-demodulator and DOC circuits. The DOC circuits sense signal dropouts and op-

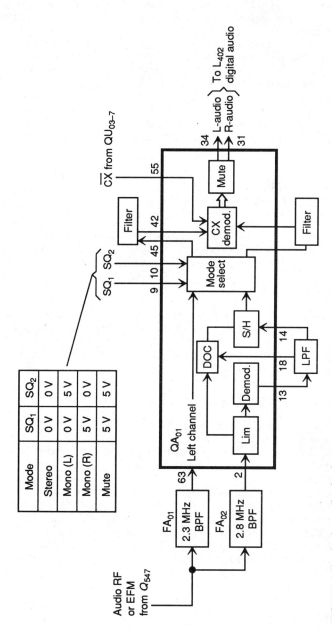

FIGURE 6.40 Analog audio circuits.

erate an S/H circuit to hold a sample of audio in the usual manner. The demodulated audio signal is applied through an external LPF and the S/H circuits to mode-select circuits.

The mode-select circuits select stereo, left-channel, right-channel, or mute according to the logic levels of the SQ_1 and SQ_2 signal (from system control QU_{03}) at pins 9 and 10 of QA_{01}. The selected audio is passed through external filters to CX noise-reduction circuits (Sec. 1.3.7). If the disc being played contains the CX noise reduction code, the $\overline{CX}$ line (from pin 7 of QU_{03}) is low, and the CX circuits in QA_{01} are activated.

The processed audio is then passed through mute circuits to the digital audio circuits. The mute circuits mute all audio during power-on and power-off.

6.17 DIGITAL AUDIO, ANALOG AND DIGITAL-SELECT, AND BILINGUAL

Figure 6.41 shows the digital-audio and analog and digital-select circuits for our CDV player, while Fig. 6.42 shows the bilingual circuits.

6.17.1 Digital Audio

The digital audio circuits are similar to those of a CD player (Chap. 5) and include demodulator Q_{301}, digital filter Q_{302}, and DAC Q_{401}. The analog output from Q_{401} is applied to the player digital and analog output jacks through analog-filter circuits Q_{403} and Q_{404} and analog-digital relay L_{402}.

6.17.2 Analog and Digital-Select

When a disc containing digital audio is played, microprocessor Q_{304} applies a low to relay driver Q_{415} which, in turn, places relay L_{402} in the digital position (DL and DR). Under these conditions, the audio appearing at the player output jacks is "digital audio" (decoded CD digital audio) in stereo analog form.

When playing a CD or CDV Single, L_{402} cannot be switched to "analog audio" since only digitally encoded audio is recorded on the CD or CDV Single. However, when playing a CDV which contains both digital audio and analog audio, the output can be switched to either digital or analog by the user.

When a CDV containing only analog audio is played, Q_{304} applies a high to Q_{415} which, in turn, places L_{402} in the analog position (AR and AL). Under these conditions, the audio appearing at the player output jacks is stereo analog audio as decoded by QA_{01} (Fig. 6.40).

6.17.3 Bilingual Circuit

The bilingual circuit allows the playback of bilingual discs as well as stereo discs. Bilingual discs are recorded with one language on the left channel and another language on the right channel. The bilingual circuit choses one of three modes: stereo, left-channel only, or right-channel only.

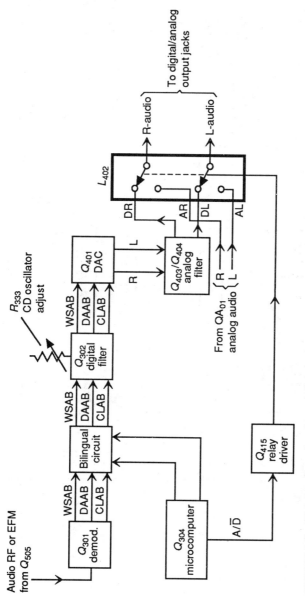

FIGURE 6.41 Digital audio and analog and digital-select circuits.

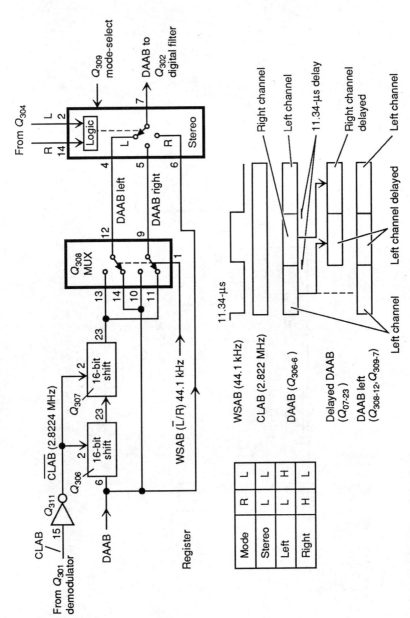

FIGURE 6.42 Bilingual circuits.

As shown in Fig. 6.42, the bilingual circuit receives CLAB (bit clock, 2.8224 MHz), DAAB (data A-chip to B-chip), and WSAB (word select A-chip to B-chip) demodulated signals from Q_{301}. (These same signals are applied to digital filter Q_{302}, as shown in Fig. 6.41.)

The CLAB signal is inverted by Q_{311} and applied to 16-bit shift registers in Q_{306} and Q_{307}. CLAB clocks the DAAB signals in and out of Q_{306} and Q_{307} to delay the DAAB data bits by 11.34 μs (as shown by the timing chart in Fig. 6.42).

Both the delayed and undelayed DAAB signals are applied to digital filter Q_{302} through multiplexer Q_{308} and mode-select switch Q_{309}. Q_{308} is switched between left and right (delayed and undelayed) DAAB signals by the WASB signal at a rate of 44.1 kHz. Q_{309} is switched between left and right by select signals from Q_{304}.

When Q_{309} is in the left position (pin 4 connected to pin 7), only the left-channel DAAB samples (SL) are passed to Q_{302}, and SL is connected to pin 14 of Q_{308} for 11.34 μs. After 11.34 μs, the WSAB signal goes high, and SL is connected to pin 13 to receive the left-channel sample a second time. Thus, instead of delivering left and right samples to Q_{302}, only left-channel samples are sent (in sequence) to Q_{302} (as shown by the timing chart).

Q_{309} can also be switched to the right position (pin 5 connected to pin 7) so that only right-channel samples are sent to Q_{302}. Likewise, Q_{309} can be switched to stereo, where pin 6 is connected to pin 7, and both channels are passed to Q_{302} in the normal manner (normal stereo operation).

6.18 HELIUM-NEON LASER CIRCUITS

Figures 6.43 and 6.44 show the circuits involved for a typical helium-neon laser. As discussed in Chaps. 1 and 2, many early-model CDV players use a helium-neon laser instead of the solid-state laser diodes now in common use. This section describes typical helium-neon laser circuits. Troubleshooting and adjustment for the helium-neon lasers are discussed in Sec. 9.12.

6.18.1 Basic Helium-Neon Laser Power-Supply Circuit

Figure 6.43a shows the basic dc power supply for the laser. The circuit is driven by a secondary winding of the power transformer T_1. Diodes D_7 and D_8 are connected as a voltage doubler. The polarities of the two diode voltages are additive, *producing about 1800 V across the capacitors*.

The resistors equalize the voltage across each capacitor and discharge the capacitors when power is turned off. C_{21} acts as a surge suppressor across the secondary winding, while C_3 and R_{17} serve as a high-pass filter to eliminate high-frequency noise which might get into the power supply.

6.18.2 Laser and Laser High-Voltage Circuit

Figure 6.43b shows the laser and laser high-voltage (HV) transformer circuit. The laser HV transformer is a totally encapsulated unit (resembling a high-voltage tripler in a TV set) mounted on the slide assembly with the laser. The circuit is

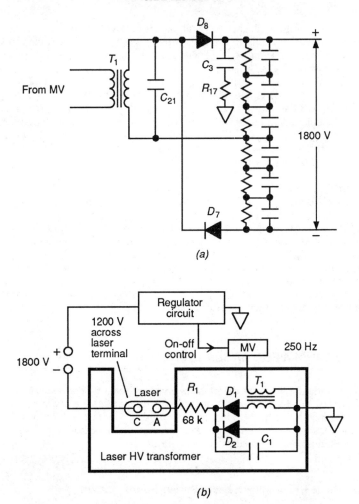

FIGURE 6.43 Basic helium-neon laser high-voltage supply circuits.

series-connected with the 1800-V source supplying current through the regulator circuit, laser, and laser transformer to ground. Note that neither side of the 1800-V power supply is connected to ground. The cathode leads from the laser passes through the laser transformer, yet no connections are made to the laser. This is done for safety purposes (with both laser leads encapsulated in the HV transformer).

Note that the 1800-V power supply does not turn the laser on. A multivibrator (MV) circuit is used to drive the primary of the HV transformer, which produces an output of about 10 kV. D_1 rectifies this voltage and C_1 charges up to about 10 kV. When this 10-kV voltage is applied across the laser anode and cathode, the laser turns on, and C_1 is discharged through R_1 to provide the initial turn-on current.

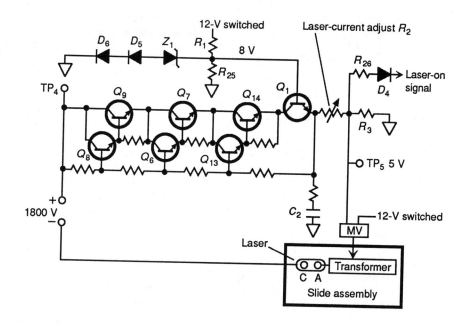

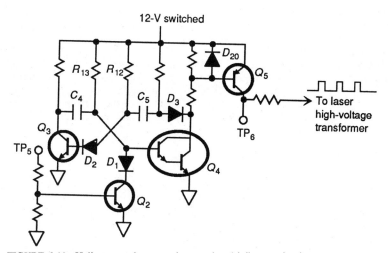

FIGURE 6.44 Helium-neon laser regulator and multivibrator circuits.

As soon as about 5 mA flows through the laser, the regulator circuit turns off the MV, and the 10-kV source is removed. The laser requires about 1200 V to maintain the 5-mA conduction current, so the 1800-V source keeps the laser on once the laser is conducting. The remaining 600 V is dropped across the regulator circuit. The 600-V level varies widely from below 100 V to over 800 V, depending on the line voltage, laser current, etc. During normal conduction of the laser, the secondary winding of the HV transformer has too much resistance to allow sufficient dc current flow. Under these conditions, the 5-mV current flows through shunt diode D_2 instead of D_1 and T_1.

6.18.3 Laser Regulator Circuit

Figure 6.44a shows the laser regulator circuit. The circuit consists of constant-current source Q_1, plus the Darlington pairs Q_8 and Q_9, Q_6 and Q_7, and Q_{13} and Q_{14}, which are used because of their high gain and resulting sensitive regulation. Transistors Q_1, Q_7, Q_9, and Q_{14} are used in series to distribute the voltage.

The +12-V switch source provides a regulated voltage at the junction of R_1 and R_{25}. Z_1 drops about 6.8 V, while D_5 and D_6 add about 1.2 V. This results in a fixed reference voltage, or bias, of about 8 V at the base of constant-current source Q_1. R_2 *is adjusted to get a fixed current of 5 mA* through the laser. R_3 is a sensing resistor that produces 5 V at TP$_5$ when the laser current is 5 mA.

The 5 V at TP$_5$ is present only when the laser is on and is coupled through D_4 to the video and servo board to serve as a "laser-on" identification signal. The 5 V are also used to turn the laser MV off (as soon as the laser turns on). The laser MV is powered from the 12-V switched source, so the circuit can never operate when the player lid is open. Capacitor C_2 charges to the emitter voltage of Q_1. When the 12-V source is removed (such as when the lid is open), the voltage on C_2 reverse-biases Q_1 and guarantees instant shutoff of the laser.

The laser is a high-impedance device and drops to about 200 V when in operation. The difference between the laser drop and the 1800 V is dropped across the regulator transistors. The 1800-V power supply can vary by several hundred volts, depending on regulation of the ac line.

The helium-neon laser must never be shorted or bridged in any matter, not even with a high-impedance voltmeter. If this rule is not observed, the series transistor can go into cascade failure. It is not necessary to bridge or short the laser since the voltage from TP$_4$ to ground is a measure of the voltage across the transistors. (The laser and HV transformer drop the difference between the TP$_4$ voltage and the 1800-V power supply.)

6.18.4 Laser Multivibrator

Figure 6.44b shows the laser MV circuits. The MV consists of Q_3 and Q_4 and is of the free-running type.

When the player lid is closed, the 12-V switched power supply turns Q_4 on through R_{13}. The collector of Q_4 goes near ground, and C_5 begins to charge toward 12 V. When the C_5 voltage reaches about 1.2 V, D_2 and Q_3 switch on. The collector of Q_3 begins to charge toward 12 V. When the C_4 voltage reaches about 1.2 V, Q_4 switches back on, and the process repeats.

Q_3 and Q_4 produce about a 250-Hz square wave that drives laser HV transformer T_1 (Fig. 6.43) through a high-current amplifier Q_5. The driving waveform can be monitored at TP_6 *but is present only for a brief instant during turn-on of the player*. As soon as the laser fires, 5 V are present at TP_5 and turn Q_2 on, shorting the base of Q_4 to ground through D_1. This disables Q_4 and turns the MV off.

A Darlington is used for Q_4 because of the high gain necessary to drive Q_5. D_2 creates a two-junction turn-on requirement for Q_3 to balance with the two-junction turn-on requirement of D_4. D_3 and D_{20} are protection diodes used to prevent positive voltage spikes at TP_6 from damaging components or upsetting circuit operation.

CHAPTER 7
MECHANICAL OPERATION, ADJUSTMENT, AND REPLACEMENT

This chapter describes operation and adjustment and replacement procedures for the mechanical sections of typical CD and CDV players. The mechanical sections are concerned mostly with loading and unloading the disc and driving the optical pickup across it. Mechanical operation of the older top-load CD players is relatively simple when compared to the more popular front-load players.

In most top-load players, the only true mechanical function (other than moving the optical pickup via a rotating arm and drive motor, as shown in Fig. 1.2*b*) is to open the disc compartment, install a disc (manually) on the turntable, and then close the disc compartment door. For these reasons, we do not go into the mechanical sections of top-load CD players here. Instead, we concentrate on the far more complex mechanical sections of front-load CD and CDV players.

A typical front-load player has three drive motors: one for the turntable, one for the optical pickup, and a third for opening and closing the disc compartment tray or door. With a typical *horizontal front-load player* (the most popular version, Sec. 5.2) the tray is opened by a drive motor, a disc is inserted (manually) into the tray, the tray and disc are moved into the player (by the drive motor), and then the disc is installed on the turntable (by the same tray drive mechanism and usually the same drive motor). Some front-load players use a fourth drive motor to clamp, or "chuck," the disc onto the turntable.

We discuss the mechanical sections for all three types of front-load players (vertical door, horizontal tray with one load and unload motor, and horizontal with a separate clamping, or chucking, motor) in this chapter. However, we concentrate on horizontal players where both the tray and clamping operations are controlled by one motor, as discussed in Chap. 5. Most CD player and CD-ROM manufacturers have adopted this version.

Since operation of the mechanical sections (drive motors, gears, etc.) are controlled by limit switches and the system-control microprocessor, the following descriptions also include diagrams and discussions of the control circuits (in simplified form). By studying the mechanical operation and circuits found here, you should have no difficulty in understanding the mechanical operations of similar CD and CDV players.

This understanding is essential for logical troubleshooting and repair, no matter what type of player is involved. For example, if you know that a particular drive motor is actuated to turn a gear in a given direction for a given mode operation, and you can see that the motor does not drive the gear in that mode, you have pinpointed a failure.

The origin of mechanical troubles may be *electronic* (no actuating signal is received from the microprocessor, or the motor may be burned out), *mechanical* (the gears may be jammed or a belt broken, etc.), or *adjustment* (limit switch is not opened or closed at the right time), but you have a starting point for troubleshooting. The descriptions given here should also help you to interpret the mechanical sections of player service literature (which are usually very good as far as adjustment procedures are concerned but often somewhat vague as to how the mechanism operates).

We also describe adjustment and replacement procedures for the mechanical sections of CD and CDV players in this chapter. Since manufacture's instructions are usually absent and to show you what typical adjustment and replacement procedures involve, we describe complete procedures for the mechanical sections, *as recommended by the manufacturer*. The procedures covered here involve the use of the special tools described in Chap. 4 and are in addition to the electrical adjustments discussed in Chaps. 8 and 9.

Remember that these specific procedures apply directly to the CD and CDV players described in this chapter. When repairing other players, you must follow manufacturer's service instructions exactly. Each type of CD or CDV player has its own adjustment points and replacement procedures. Using the examples described here, you should be able to relate the procedures to a similar set of adjustment points and replacement procedures on most similar players. Where it is not obvious, we also describe the purpose of the procedure.

7.1 CD VERTICAL FRONT-LOAD MECHANICAL SECTION

Figure 7.1 shows the location and principle mechanical components of a typical front-load CD player with a vertical door. Figure 7.1 also shows the associated wiring.

Note that most of the components are mounted on a *unit base* to which the vertical *door assembly* (loading mechanism) is hinged (at the bottom). The door is attached to the base by *L-arms* and is opened and closed by a *crank arm* driven by a *gear motor*. The door is supported by *springs* (A and B), and *rollers* are mounted on arms which are part of the door assembly.

The door is opened by the gear motor, a disc is installed (manually), and the door is closed by the motor. The disc is pressed against the *turntable motor assembly* by a disc clamp assembly on the door. The turntable motor is operated by a turntable-drive servo (Chap. 5) to spin the disc at the appropriate speed.

As shown in Fig. 7.1*b*, the gear motor for the door assembly receives open and close drive signals from the system-control microprocessor through gates and transistors. In turn, the microprocessor receives control signals from the front-panel Open/Close switch. The microprocessor also receives indicator signals from the door-open (LIDO) microswitch and the door-close (LIDC) microswitch. The LIDO and LIDC switches are positioned and adjusted to actuate when the door has reached the correct open and close limits and thus cut off the door motor through operation of the microprocessor.

MECHANICS, ADJUSTMENT, AND REPLACEMENT 7.3

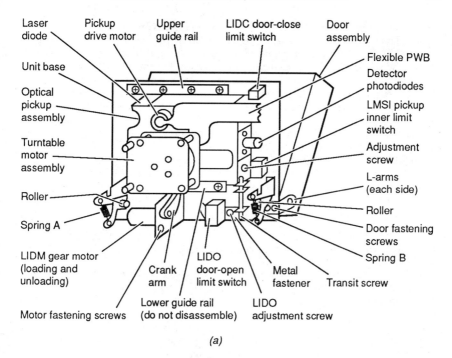

(a)

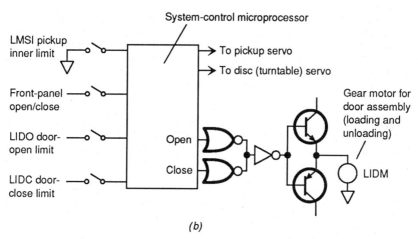

(b)

FIGURE 7.1 Typical front-load CD player with vertical door.

The *optical pickup assembly* is secured to the unit base by upper and lower guide rails and is driven across the disc by a motor (which is part of the pickup assembly). The motor is operated by a servo, as described in Chap. 5. The pickup motor is connected to the pickup drive gears by means of a belt. The belt can be replaced through a cutout on the unit base, *without removing either the pickup or the motor* (on most players). The cutout is normally covered by a press-out lid, or cover. The optical pickup is held in place when the *transit screw* is tightened.

The entire pickup can be replaced as an assembly, or it can be replaced separately (we describe both procedures in this section). Some manufacturers supply a few additional mechanical parts (gears, etc.) for replacement on the pickup assembly. Always consult the service literature regarding such replacement parts. However, as a practical matter, you will probably replace the belt, the motor, or the entire pickup assembly. In any event, you should *never* attempt to replace the laser diode or the detector photodiodes unless specifically directed to do so by the service literature (which is not likely).

As shown in Fig. 7.1*b*, the system-control microprocessor receives a signal from the pickup inner-limit (LMSI) microswitch. The LMSI switch is positioned and adjusted to actuate when the pickup assembly has reached the inner limit (start) of the disc. The LMSI signal cuts off the pickup motor through operation of the microprocessor and pickup servo.

7.1.1 CD Mechanical Section Servicing Precautions

Here are some precautions you should observe before performing any adjustment or replacements on the mechanical sections of a vertical front-load CD player. These precautions are *in addition* to any precautions found in the player literature.

Laser Safety. Although you will probably be performing all adjustments and replacement of parts in the mechanical section with the power off (except possibly to test for proper adjustment of the limit switches), *remember that the laser diode, on a vertical front-load player, can operate even when the door is open.*

Disassembly. The mechanical section is precision engineered and critically adjusted at the factory. Do not disassemble any part of the player beyond that point absolutely necessary to gain access or replace parts. This precaution applies especially to the pickup assembly (lens actuator, laser diode, optical parts, detector photodiodes) and to the *lower guide rail* (Fig. 7.1*a*). The lower guide rail should never be removed from the unit base since the rail serves as a reference standard for the pickup mount dimensions (as we discuss later in this section).

Aluminum Parts. Most of the mechanical parts are made of aluminum. Be careful not to overtighten screws, and do not scratch or bend any parts by exerting excessive force. (It is very easy to strip the threads in the unit base.)

Flexible Wiring. The pickup assembly has flexible printed wiring, since the pickup must move across the disc. While the wiring is sufficiently strong to withstand constant movement of the pickup, the wiring is not designed to be bent or twisted during service. Remember that if you break even one lead in the wiring, the entire pickup must be replaced.

Dirt in Moving Parts. Be especially careful not to get any dirt or other foreign matter in the guide rails or the turntable assembly. The pickup assembly must be completely free to move within the rails, and the turntable must be capable of spinning freely.

Objective Lens. If the player has been in service for any length of time, always check the objective lens for dirt, dust, smuges, and so on (but keep your fingers away from the lens while doing so). Use a clean, dry cloth to clean the lens. If the objective lens is dirty, the EFM signal can be weak. If you find a weak EFM signal during troubleshooting (Chap. 8), always make a quick check of the objective lens.

Wiring. Be careful to route all wiring to the original position. It is particularly important to keep any wiring from being jammed by the pickup assembly or the door open and close mechanism, since both of these sections are subject to repeated movement.

Adjustment after Replacement. Always perform the associated electrical adjustment procedures after replacement of any mechanical parts. For example, if the turntable motor or assembly is replaced, the Hall-element gain and offset adjustments should be checked (as described in Chap. 8). If the pickup assembly is replaced, check the laser-diode output and tracking and focus adjustments.

Lens Actuator Quick Check. If you suspect that the lens actuator on the pickup assembly is defective, it is possible to make a quick check of the actuator without removing the entire pickup assembly for replacement. Simply measure the resistance of focus and tracking coils with an ohmmeter, as shown in Fig. 7.2.

Typically, the focus coil is about 20 Ω, while the tracking coil is 4 Ω. The actual resistance depends on the particular assembly. However, if you get an open or short indication, or a resistance that is drastically different from these values, the actuator is suspect. In some players, you can see a slight movement of the actuator when the ohmmeter is connected to the coils. This usually indicates that the actuator is good.

Transit Screw. In most cases, the transit screw should be in place when performing any service procedures. This prevents the pickup assembly from moving back and forth when the player is moved about the service bench. Of course, you must loosen the transit screw for certain checks.

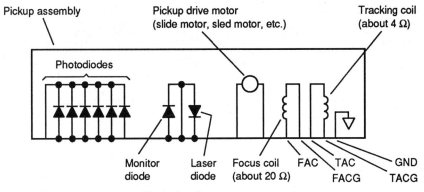

FIGURE 7.2 Lens actuator quick-check points.

7.1.2 Gaining Access to the Components

This section describes the basic procedures to open the player for service. Remember that these procedures are "typical" for vertical-door players. Always follow any disassembly procedures found in the service literature. Figures 7.3 and 7.4 illustrate the disassembly procedures.

Figure 7.3a shows the procedure for taking off the cover. Slide off the back cover after disengaging the screws (1) and (2).

Figure 7.3b shows the procedure for taking off front panels A and B. Pull out the nylon rivets holding the front panels. Use a pointed object to remove the rivets.

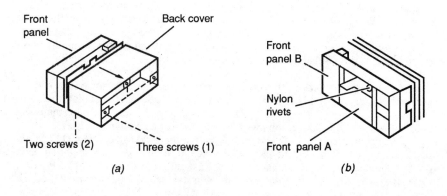

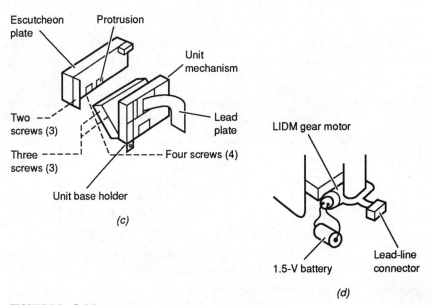

FIGURE 7.3 Gaining access to vertical-door components.

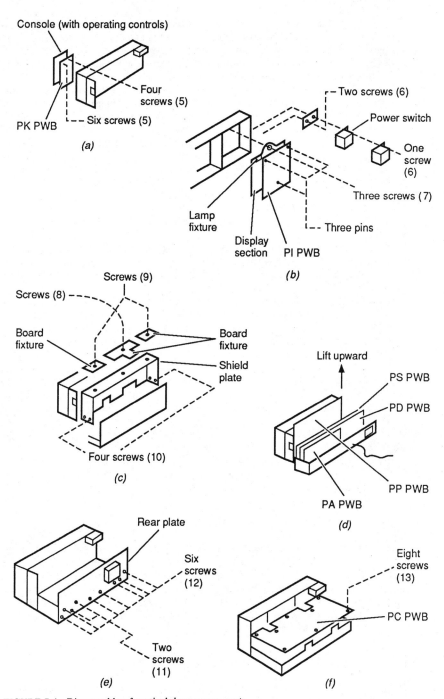

FIGURE 7.4 Disassembly of vertical-door components.

Figure 7.3c shows the procedures for removing the unit mechanism. After taking off the cover, remove screws (3), which hold the escutcheon plate. Switch on the power, push the door Open/Close button, and open the door. Then switch off the power and unplug the power cord. Next, after removing the escutcheon plate, remove the lead plate and disengage screws (4).

As shown in Fig. 7.3c, there is a pressed protrusion on the lower part of the escutcheon. When dismounting or mounting the unit mechanism, be careful to cover the protrusion with a piece of thick paper (such as drawing paper) so as not to damage the lower part of the unit mechanism.

If the door does not open after pressing the door Open/Close button, use a 1.5-V battery to operate the door motor, as shown in Fig. 7.3d. After removing the lead connector, connect a 1.5-V battery directly to the motor terminals as shown. The door can be closed by reversing the battery terminals. (This same procedure can be used as a quick check of the door motor, if you are having door open and close problems).

Avoid lifting or moving the unit mechanism by the unit-base holder (which can easily slip out of place).

Figure 7.4a shows the procedure for removing the console. Disengage screws (5) and move the console forward as shown. Remember that the console need not be removed unless operating controls and the associated printed wiring are to be replaced.

Figure 7.4b shows the procedure for removing the power switch and display section. Disengage screws (6) to remove the power switch. Disengage screws (7) to remove the display section. Remember that the display section need not be removed unless the lamp fixture, LEDs, and so on and the associated printed-wiring board are to be replaced.

Figures 7.4c through 7.4f shows the procedures for removing the various printed-wiring boards (PWB). Disengage screws (8) and the board fixture to remove the PA PWB. Disengage screws (8) and (9) and the board fixture to remove the PP PWB. Disengage screws (10), the board fixture, and the shield plate to remove the PD and PS PWBs. After removing the escutcheon plate and the PA, PP, PD, and PS PWBs, disengage screws (11) and (12) and then remove the rear plate (Fig. 7.4e). Disengage screws (13) and remove the PC PWB (Fig. 7.4f).

After removing the console (Fig. 7.4a), disengage screws (5) and remove the PK PWB. After removing the display section (Fig. 7.4b) and lamp fixture, remove the three pins and then remove the PI PWB.

7.1.3 Changing the Pickup Drive Belt

Figure 7.5 shows the basic procedures required to change the pickup assembly drive belt. Switch on the power. Press the door Open/Close button to open the door (or use a battery, as shown in Fig. 7.3d). Switch off the power and turn the left-hand side of the unit face down, as shown in Fig. 7.5. Remove the belt-access cover, or lid, from the unit base by moving the cover in the direction of the arrows. Use tweezers to remove the old belt and to install a new belt over the motor axle and pulley, as shown. Replace the access cover and close the door.

7.1.4 Removing the Door (Loading Mechanism) Assembly

This section describes the basic procedures to remove the door assembly from the unit base. This should be done *only* if the door assembly must be replaced.

MECHANICS, ADJUSTMENT, AND REPLACEMENT

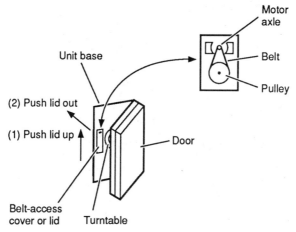

FIGURE 7.5 Changing vertical-door pickup-assembly drive belt.

Before performing any repairs or replacement, switch on the power, set the player in the Stop mode, and fasten the transit screw (to prevent the pickup from sliding back and forth).

Remove springs A and B from the roller shaft (Fig. 7.1). Remove the unit mechanism from the main chassis (Fig. 7.3c). Close the door by connecting the motor terminals to the main chassis or by applying 1.5 V from the battery (Fig. 7.3d).

Remove the four door-fastening screws from the L-arms, as shown in Fig. 7.1. Then remove the door assembly using a hex wrench key for the gear-motor crank arm.

7.1.5 Reinstalling the Door (Loading Mechanism) Assembly

This section describes the basic procedure to reinstall the door assembly on the unit base. Again, the transit screw should be in place. Figure 7.6 shows the procedures.

Before installing the door assembly, loosen (but do not remove) the gear-motor crankshaft fastening screw. Insert the door assembly and temporarily fasten the L-arm with the four door-fastening screws, as shown in Fig. 7.6a. Set the door-assembly aligning jig into the disc-loading area. Position the door-assembly aligning spacer as shown in Fig. 7.6a. Note that this operation cannot be performed if the gear-motor crank-arm fastening screws are tightened.

While pushing the door-assembly aligning jig, turn the rotor section of the turntable motor by hand, as shown in Fig. 7.6b. Confirm that the turntable does not turn easily (because of the pressure from the jig). Keeping this condition, tighten the door-fastening screws.

Remove the door-assembly aligning jig and the door-assembly aligning spacer. Insert a check disc, as shown in Fig. 7.6c. With the door closed, turn the rotor section of the turntable motor and confirm that the disc turns smoothly without contacting any part of the loading area.

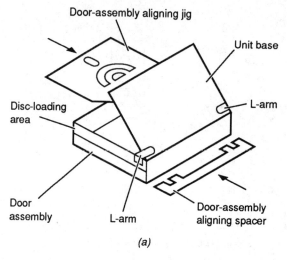
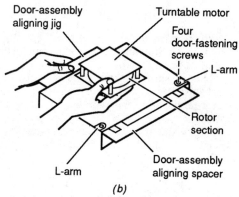
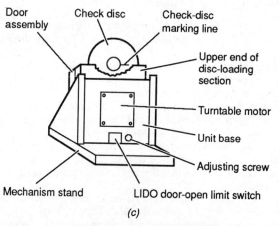

FIGURE 7.6 Reinstalling the vertical-door assembly.

MECHANICS, ADJUSTMENT, AND REPLACEMENT   7.11

Set the assembled unit-mechanism assembly on a mechanism stand (Fig. 7.6c) and plug in the appropriate connectors to the PS and PK PWBs. Switch on power and open the door.

Open and close the door several times after inserting the check disc. Adjust the position of the LIDO microswitch (door-open limit) so that the check-disc marking line is below the upper end of the disc-loading section when the door is open, as shown in Fig. 7.6c.

7.1.6 Replacement of Gear Motor (for Door Open and Close)

This section describes the basic procedure to replace the door motor without complete disassembly of the unit. Refer to Figs. 7.1 and 7.7a for the location of parts. The transit screw should be in place.

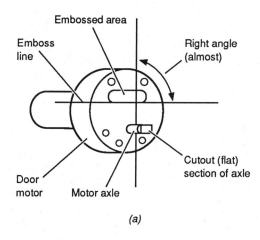

(a)

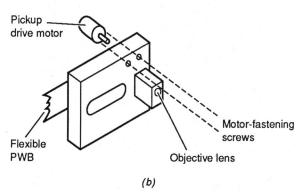

(b)

FIGURE 7.7 Replacement of vertical-door gear motor and pickup drive motor.

Loosen the gear-motor crank-fastening hex screw, using a hex wrench. Then remove the two motor-fastening screws. Disconnect the wiring. Reverse the procedure to install the motor.

It is easier to replace the door motor if you follow the guidelines shown in Fig. 7.7a. The cutout section of the motor axle must be aligned when installing the motor. Correct alignment is achieved when the cutout (or flat) section of the axle makes a right angle (almost) with the emboss line on the motor face. Use a 1.5-V battery to rotate the motor axle as necessary.

7.1.7 Replacement of Pickup-Drive Motor

This section describes the basic procedure to replace the pickup motor without complete disassembly of the unit. Refer to Figs. 7.1 and 7.7b for the location of parts. The transit screw should be in place.

Remove the lid of the belt-replacement access cutout (Fig. 7.5). Unsolder the motor connections from the flexible PWB. Remove the motor connections from the flexible PWB on the rear. Remove the two motor-fastening screws. Remove the motor. Reverse the procedure to install the motor. Make certain to replace the belt as described in Sec. 7.1.3, after the motor is installed.

7.1.8 Replacement of the Turntable Motor

This section describes the basic procedure to replace the turntable motor without complete disassembly of the unit. Refer to Figs. 7.1 and 7.8 for the location of parts. It is necessary to loosen the transit screw during this procedure.

Before replacing the turntable motor, try rotating the turntable by hand. The rotor section should turn smoothly. If it does not, the rotor portion of the motor

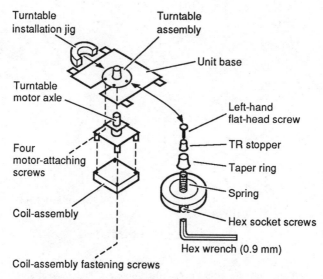

FIGURE 7.8 Replacement of vertical-door turntable motor.

must be replaced. If the rotor turns smoothly but the motor does not run when power is applied, it is possible that the coil assembly is defective and can be replaced separately (on some but not all players). Also, some turntable motors require lubrication on the axle or drive shaft during replacement. Always check the service literature.

Start by removing the door assembly as described in Sec. 7.1.4. (In most vertical-door players, you cannot replace the turntable and motor without removing the door. However, you can replace the coil assembly without removing it.)

Remove the left-hand flat-head screw and the TR stopper shown in Fig. 7.8. Loosen the two hex screws, and remove the turntable assembly. Be careful not to lose the screw, TR stopper, spring, or taper ring.

Now loosen the transit screw and move the pickup toward the outside. Reach through the lens actuator access hole and remove the four turntable motor-fastening screws. Position the pickup at the center of the guide rail, turn the turntable motor 90°, and then remove the motor. Reverse the procedure to install the turntable assembly and motor.

The turntable height must be properly set during installation. Use a turntable installation jig, as shown in Fig. 7.8. Slide the turntable assembly over the motor axle or drive shaft, with the jig positioned between the unit base and turntable-assembly bottom. Tighten the two hex screws. Remove the jig (which should be snugly in place, but not binding), and check that the turntable rotates freely.

If the coil assembly is available as a separate replacement part (typical for many players), simply remove the four screws and disconnect the wiring (unplug the connector).

7.1.9 Installing and Adjusting the Microswitches

This section describes the basic procedure to install and adjust the three mechanical-section microswitches. Refer to Figs. 7.1 and 7.9 for the location of parts. The transit screw should be in place.

Install the LIDC door-close limit switch so that the slider of the switch and the inside of the hole on the unit base *are not in contact*.

Turn the adjustment screw (Fig. 7.9) for the LMSI pickup inner-limit

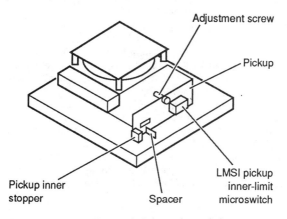

FIGURE 7.9 Installing vertical-door microswitches.

microswitch so that switch operation is triggered when the gap between the pickup inner stopper and the pickup is 1 mm. Use a spacer for this adjustment.

Adjust the LIDO door-open limit switch as described in Sec. 7.1.5 and shown in Fig. 7.6c. Loosen the LIDO adjustment screw (Fig. 7.1) and slide the LIDO switch and metal fastener as necessary. Tighten the adjustment screw once proper adjustment is obtained. The check-disc marking line should be below the upper end of the disc-loading section when the door is fully open (the LIDO switch actuated).

7.1.10 Removing the Pickup Assembly

This section describes the basic procedure to remove the pickup assembly from the unit base. This should be done only when you are certain the pickup assembly is defective. Be sure to observe all precautions regarding static discharge described in Chap. 4. It is necessary to loosen the transit screw during this procedure. Refer to Figs. 7.1 and 7.10 for the location of parts.

First remove the turntable motor, as described in Sec. 7.1.8. Then remove the four fastening screws for the *upper guide rail* (but not the lower guide rail). Disconnect the wiring (flexible PWB) and lift the pickup assembly from the unit base. Be careful not to lose the upper and lower crossed-roller assemblies (positioned between the guide rails and pickup assembly).

7.1.11 Reinstalling the Pickup Assembly

This section describes the basic procedure to reinstall the pickup assembly on the unit base. Refer to Figs. 7.1 and 7.10 for the location of parts. It is necessary to loosen the transit screw during this procedure.

Install the lower crossed-roller assembly into the V-shaped groove of the lower guide rail (making the right and left spacing equal). If specified by the service literature, lubricate the upper and lower crossed-roller assemblies (typically with silicone grease, such as HIVAC-G).

Set the pickup assembly on the lower guide rail while aligning the outer side of the pickup with the outer side of the guide rail.

Install the upper crossed-roller assembly into the V-shaped groove on top of the pickup assembly (Fig. 7.10). Make the right and left spacing equal. Align the upper guide rail and temporarily tighten the four fastening screws so that the pickup assembly can still be moved freely.

Remove the lid of the belt-replacement access cutout in the unit base (Fig. 7.5). Squeeze the top and bottom edges of the upper guide rail at the center with thumb and forefinger, as shown in Fig. 7.10. Apply force at an angle of about 45° to the upper guide rail face. In this condition, first tighten the two inner screws and then the two outer screws. Move the pickup assembly and check that the pickup moves smoothly without binding (too tight) or chattering (too loose). Tighten the transit screw to lock the pickup assembly in place.

MECHANICS, ADJUSTMENT, AND REPLACEMENT   7.15

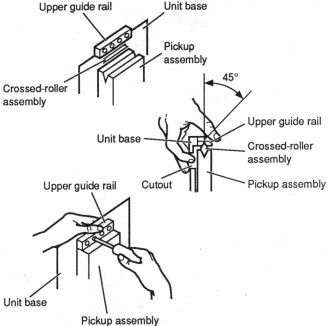

FIGURE 7.10 Reinstalling the vertical-door pickup assembly.

7.2 CD HORIZONTAL FRONT-LOAD MECHANICAL SECTION (SINGLE LOAD MOTOR)

Figure 7.11 shows the major mechanical components for a typical front-load CD player with a horizontal tray operated by a single open, close, load, and unload motor. Figure 7.11 also shows the associated wiring. Note that most of the components are part of a unit mechanism secured to the mainframe by two rails.

The tray is moved out of the player front panel by the loading motor (LIDM). This action also raises the *clamp*, or *chuck*. (Compare this to the functions shown in Fig. 5.2a.) A disc is installed manually in the tray, and the tray is pulled within the player by the loading motor. This action also lowers the clamp, or chuck, so that the disc is pressed against the *turntable motor assembly*. The turntable motor is operated by a turntable drive motor (Chap. 5) to spin the disc at the appropriate speed. In most players, the *coil assembly* can be separated from the turntable motor and replaced as a separate component.

The two *laser safety interlock* switches (one for the cover and one for the tray) are connected in series. Both switches must be actuated (tray fully within the player, and the player cover in place) before the laser-drive circuits can operate. (This is similar but not identical to the laser-drive circuits discussed in Chap. 5.)

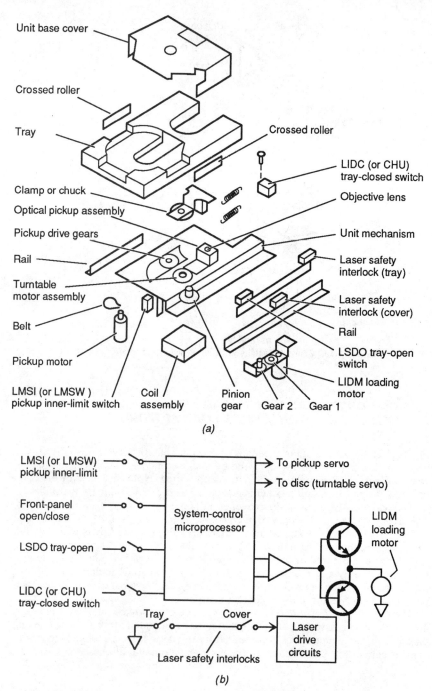

FIGURE 7.11 Typical front-load CD player with horizontal tray (single load motor).

MECHANICS, ADJUSTMENT, AND REPLACEMENT   7.17

The loading motor LIDM receives open and close drive signals from the system-control microprocessor through gates and transistors. In turn, the microprocessor receives control signals from the front-panel Open/Close switch. The microprocessor also receives indicator signals from the tray-open (LSDO) and the tray-closed (LIDC) switches. Note that the LIDC switch is actuated when the tray is in and the clamp, or chuck, is in the fully down position. (LIDC is identified as the chuck, or CHU, switch in some literature.)

The LSDO switch is positioned and adjusted to actuate when the tray has reached the correct open limit and thus cuts off the loading motor through operation of the system-control microprocessor. No matter what it is called, the LIDC, or CHU, switch actuates when the tray is in and the clamp is fully down, thus cutting off the loading motor.

The *optical pickup assembly* is driven across the disc by the pickup motor, which is part of the unit mechanism. The pickup motor is operated by a servo, as discussed in Chap. 5. The pickup motor is connected to the *pickup drive gears* by a belt. The belt can be replaced when the player cover and the *unit base cover* are removed, without removing either the pickup or the motor.

The system-control microprocessor receives a signal from the pickup inner-limit (LMSI) switch. This switch is positioned and adjusted to actuate when the pickup assembly has reached the inner limit (start) of the disc. The LMSI signal cuts off the pickup motor through operation of the microprocessor and the pickup servo. (The LMSI switch is called the LMSW switch in some literature, such as in Fig. 5.2*b*.)

The entire unit mechanism can be replaced as an assembly (in most players). Some manufacturers also recommend replacement of the motors and limit switches (and they describe the procedures for replacement and adjustment in their service literature). As a practical matter, never disassemble the unit mechanism beyond that point necessary to replace or adjust a given part. Never make any adjustments unless the troubleshooting leads you to believe that adjustment is required.

7.2.1 Operation of the Disc Loading and Unloading Mechanism

The following paragraphs describe how the tray is moved in and out and how the clamp, or chuck, is lowered and raised by the single loading motor. Remember that this applies to a specific horizontal front-load CD player (with single motor) but is generally correct for most similar players (and represents the most popular configuration for CD players).

Loading and Clamping the Disc. Figure 7.12*a* shows the sequence for loading and clamping (chucking) the disc in a horizontal front-load CD player. As shown, the operation can be divided into three basic steps (tray loading, moving the disc, and clamping). The *loading motor* is used as the driving force for all three steps. The loading-motor rotation is reduced through gears, then converted into horizontal motion of the *rack* under control of a *latch* and link mechanism. The rack is divided into two sections. One moves as a body with the tray; the other moves separately (to move the disc up and down and to provide clamping).

The tray is loaded 150 mm into the player in about 2 s. The rack is then fixed to the tray with a latch. When the tray moves fully into the player, the latch is disengaged, the rack then slides, the *lifters* are lowered, and the disc is moved down. This takes about 0.15 s. While the rack is moving 35 mm, the *clamp bar* is

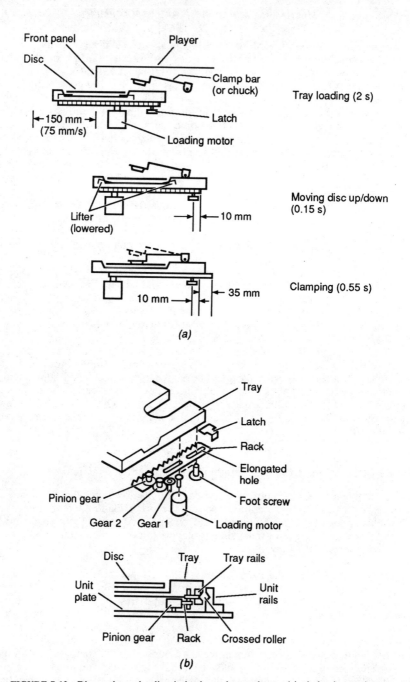

FIGURE 7.12 Disc and tray loading in horizontal tray players (single load motor).

MECHANICS, ADJUSTMENT, AND REPLACEMENT   7.19

lowered to clamp the disc. The time required for this operation is about 0.55 s. When play is complete, the operation is reversed (the clamp bar is raised, the lifters raise the disc, and the tray moves outside the player).

The loading motor is turned off by the LSDO switch (Fig. 7.11) when the tray reaches the fully open, or extended, position. The loading motor is also turned off by the LIDC switch when the tray is fully in and the clamp is fully down. The time between fully open and fully closed is about 2.75 s. If the system-control microprocessor senses that the time between open (LIDO actuated) and close (LIDC actuated) is greater than about 5 s (say that you have caught your finger in the tray), the microprocessor causes the loading motor to open the tray (move the tray to the fully open position).

Loading the Tray.   Figure 7.12*b* shows details of tray loading. The tray is mounted on the *tray rails* (left and right) and *unit rails* (left and right). The tray rails are part of the tray, while the unit rails are part of the *unit plate*. Power from the *loading motor* is applied to the rack through *gear 1, gear 2,* and the *pinion gear*, causing the tray to move in and out in a horizontal plane. The rack can move about 45 mm separately from the tray. However, during tray loading, the *latch* works as a stopper so that the rack and tray move as a single unit or body.

Sliding the Rack.   Figure 7.13*a* shows the relationship of the rack and tray. One section of the rack slides independently of the tray to move the disc and to move the clamp up and down. During closing, as soon as the tray comes to the close position, the latch is fed into the *hole* of the *latch guide* and rotates in the direction of the A arrow, permitting the rack to slide in the direction of the B arrow.

While the tray is opening, the rack first moves about 45 mm, and the *R part* of the rack contacts the *foot screw* to open the tray. At this time, the latch (which has moved into the hole of the latch guide) moves in the direction of the C arrow so that the rack and tray move as a body.

Moving the Disc up and Down.   Figure 7.13*b* shows the details of disc up and down movement. The disc is moved up and down by the elevation (rotation) of the four *lifters* (which are part of the tray). The lifters are coupled to the *lifter cam assembly*, which is composed of a *lifter cam* and two *cam levers* (left and right). The two cam levers are moved (around two *fulcrums*) by the lifter cam. In turn, the lifter cam is moved by the *lifter spring* (in the down direction) and by the *rack contact* at point E (in the up direction). Force is normally applied to the lifters in the up direction by the *up springs*.

During load, when the disc is to be moved down, the rack moves together with the tray until the tray is within the player. During this time, the lifter-cam assembly does not contact the lifters, which are held up by the lifter spring. When the tray reaches the close position, the rack starts moving backward, point E is separated, but the lifter cam is also moved backward (like the rack) by the force of the lifter spring, thus moving the lifters (and disc) down. The rear lifters are operated directly by the lifter cam, while the front lifters are moved to the down position by the left and right cam levers.

During unload, when the disc is to be moved up, the force from the loading motor which moves the rack forward is received at point E. This moves the lifter-cam assembly forward to separate the lifter cam from the lifters. The rear lifters are separated from the lifter cam, while the front lifters are released by the left and right cam levers. The lifters are moved to the up position by the up springs.

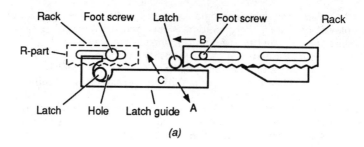

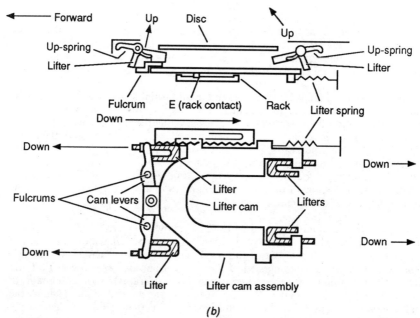

FIGURE 7.13 Details of rack, tray, and disc in horizontal tray players (single load motor).

Clamping the Disc. Figure 7.14 shows details of disc clamping. The *disc* is clamped onto the *turntable* by the *clamp bar*, which uses the horizontal movement of the *rack* as a drive force. The rack moves about 35 mm after the disc is moved down. The rack has a slope A, which contacts the boss of the C cam. This rotates the C cam in the direction of the P arrow. The C cam motion is coupled through the C cam springs to the C arm, which moves the clamp bar over the disc.

During load, when the rack moves backward and the disc is clamped, slope A rotates the C cam in the direction of the P arrow. At this time, the C arm is rotated in the direction of the S arrow by the force of the C cam spring (1), thus clamping the disc onto the turntable.

During unload, when the rack moves forward and the disc is unclamped, the

MECHANICS, ADJUSTMENT, AND REPLACEMENT 7.21

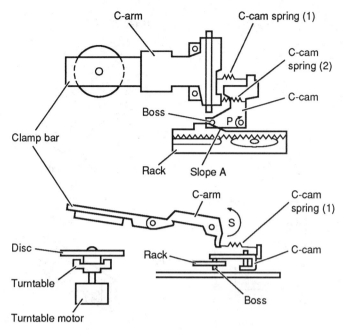

FIGURE 7.14 Details of disc clamping in horizontal tray players (single load motor).

boss of the C cam is moved along slope A by the force of C cam spring (2), turning the clamp bar up to unclamp the disc from the turntable.

7.2.2 Mechanical Section Servicing Precautions

All precautions described in Sec. 7.1.1 generally apply to horizontal front-load CD players. Here are some additional precautions to consider.

In many horizontal front-load CD players, the laser diode does not receive drive signals (Fig. 7.11b) unless the tray is fully retracted (to actuate the tray laser safety interlock switch) and the player cover is in place (to actuate the cover interlock switch). In players with circuits such as shown on Fig. 5.2b, only one laser safety interlock (chuck switch S_3) is used. With either configuration, you must override these interlocks during service. Try to avoid this. If you must override the interlocks, *always avoid direct exposure to the laser beam. Never look directly into the objective lens.*

7.2.3 Gaining Access to the Components

This section describes the basic procedures to open the player for service. Remember that these procedures are "typical" for horizontal CD players. Always follow any disassembly procedures found in the service literature. Figure 7.15 shows the components involved.

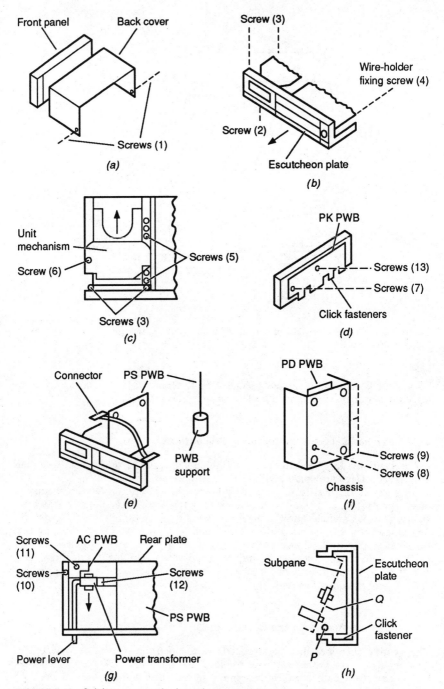

FIGURE 7.15 Gaining access to horizontal tray components.

MECHANICS, ADJUSTMENT, AND REPLACEMENT 7.23

Figure 7.15a shows the procedure for taking off the cover. Slide off the back cover after disengaging the screws (1).

Figure 7.15b shows the procedure for removing the escutcheon plate. After removing the cover, disengage screws (2) and (3), and wire-holder fixing screw (4). Pull the escutcheon plate to the front.

Figure 7.15c shows the procedure for removing the unit mechanism. After removing the escutcheon plate, disengage screws (5) and (6). Slide the unit mechanism backward (in the direction of the arrow), and disconnect the unit lead wires from the unit mechanism.

Figures 7.15d through 7.15g show the procedures for removing the various PWBs and the power transformer. After removing the escutcheon plate, disengage screws (7), and then take off the lower click fasteners to remove the PK PWB. Hold the upper end of the plastic PWB support (Fig. 7.15e) at both sides, and remove the support from the PS PWB. It is convenient to insert the removed PS PWB into the slit of the support and keep it there during repair.

Disengage the screws that hold the chassis of the PWB support. Stand the player with the unit mechanism facing down (Fig. 7.15f). Disengage screws (8) and (9) and remove the chassis. After removing the unit mechanism, pull the power lever in the direction of the arrow to remove the lever. Disengage screws (10) and (11) and remove the ac PWB in the direction of the rear plate. After removing both the unit mechanism and the ac PWB, disengage screws (12) and remove the power transformer.

Figure 7.15h shows the procedure for removing the output volume control and headphone jack. After removing the escutcheon plate, lift the lower subpanel click fasteners in the direction of the P arrow. Rotate the subpanel in the direction of the Q arrow, engaging the upper part, and remove the subpanel. Then remove the output volume control and headphone jack attached to the subpanel.

After removing the escutcheon plate, take off the PK PWB and disengage the screw (13) (Fig. 7.15d). The control assembly can then be removed.

7.2.4 Changing the Pickup Drive Belt

Figure 7.16a shows the basic procedures required to change the pickup assembly drive belt. Disengage screws (14) and remove the unit base cover. Disconnect the LED connector. Turn the power on. Press the disc Open/Close switch to open the disc tray. Use tweezers to remove the old belt and to install a new one. To mount the new belt, first install the belt on the pulley, and then on the motor shaft, using tweezers. Be careful not to touch the objective lens during this process.

7.2.5 Adjusting the Loading Mechanism

Figure 7.16b through 7.16f show the procedures for adjusting the disc-loading mechanism, including the open and close limit switches and the laser safety interlock switches.

Adjusting the Tray-Closing Position. Close the tray, as shown in Fig. 7.16b. Turn the *right adjusting screw* so that the screw just contacts the tray. Then turn the right adjusting screw another half turn (about 0.25 mm). Repeat the same proce-

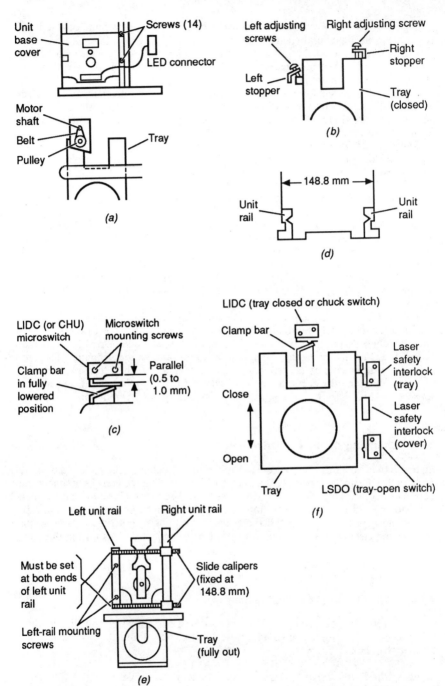

FIGURE 7.16 Changing the pickup drive belt and adjusting the loading mechanism in horizontal tray players.

dure for the left adjusting screw. Apply a sealer to the adjusting screws to prevent them from loosening.

Adjusting the Tray-Closed Microswitch Position. Adjust the LIDC microswitch (also called the chuck, or CHU, switch in some literature) so that the switch is actuated when the clamp bar is fully lowered, using the dimensions of Fig. 7.16c as a guide. Tighten the microswitch-mounting screws in this position.

Adjusting the Tray-Mounting Unit Rails. The tray-mounting unit rails must be perfectly parallel, as shown in Figs. 7.16d and 7.16e. Remove the player cover and unit-base cover. Pull out the tray. Loosen (but do not remove) the left rail mounting screws. Fix slide calipers at 148.8 mm (or whatever dimension is specified in the service literature). Secure the calipers on the rails as shown, and check that the rails are parallel.

Repeat this two or three times for best accuracy. Tighten the mounting screws in this position. Now move the tray back and forth. The tray must not bind at any point between fully open to fully closed. Equally important, there must be no chattering or vibration when the tray is moved.

Adjusting the Loading-Mechanism Switches. Figure 7.16f shows the relationship of the laser-interlock switches and the loading-mechanism limit switches to the tray and clamp bar. As shown in Fig. 7.16c, the LIDC (tray-closed, or chuck) switch is actuated (closed) by the tray when the tray is open.

The laser tray-open switch is closed only when the tray is fully in, and it opens when the tray is in *any other position* (moving or stopped). The laser cover-open switch is closed only when the player cover is in place, and it opens when the cover is removed.

As long as all of these conditions are met, the actual mounting dimensions for the switches are not critical (with a possible exception of the LIDC dimensions, Fig. 7.16c). Make certain to tighten the switch-mounting screws once you are certain that the switches actuate properly. (This same condition is true for the LMSI pickup inner-limit switch shown in Fig. 7.11.)

7.2.6 Further Disassembly of the Unit Mechanism

We do not go into further disassembly of the unit mechanism here. Typically, you can remove and reinstall the *tray mechanism, disc clamp and springs, turntable-motor coil assembly* (and possibly the *turntable motor* and *turntable assembly*), *pickup assembly, pickup motor*, and *loading motor*. Figure 7.11 shows the relationship of these components in a typical horizontal front-load CD player (the most popular model).

The procedures required for disassembly and reassembly can differ greatly from one CD player to another. You must follow the procedures and study the exploded-view illustrations (if any) found in the service literature for the particular player you are servicing. The procedures we have covered thus far in this section can serve as a guide to understanding the service literature.

7.3 CD HORIZONTAL FRONT-LOAD MECHANICAL SECTION (TWO MOTORS)

Some horizontal front-load CD players have a separate motor to operate the disc clamp (or chucking arm, as it may be called). Figure 7.17 shows such an arrangement, as well as the associated wiring. Before we get into chucking-arm operation, let us consider the sequence of the loading process.

As shown in Fig. 7.17b, the loading motor M_1 and chucking motor M_2 are driven with the amplifier IC_{304} output signals (at pin 16 of IC_{304}). The load-out and load-in signals are taken from pins 19 and 20 of system-control microprocessor IC_{102}. Switches are used to connect M_1 and M_2 or to make the motors ready for operation or to indicate the status of the mechanical components.

7.3.1 Tray-Open Sequence

The following sequence of operation occurs from the disc-tray closed state to the disc-tray open state.

When Open/Close switch S_{903} is operated, the output at pin 19 of IC_{102} goes high (about +3.2 V). Because of the inversion at pin 12 of IC_{304}, the amplified output at pin 16 of IC_{304} goes low (about −9 V).

The chucking motor M_2 is supplied power through the chucking and loading switch S_{905} (which is in the chucking position when the tray is fully in) and the disc-tray open and close switch S_{904} (which is closed when the tray is fully in). The chucking motor M_2 starts to rotate, and the chucking arm moves up to unclamp the disc (Sec. 7.3.3).

When the disc is unclamped (chucking is released), S_{905} switches to the loading position. This removes power from the chucking motor M_2 and applies power to the loading motor M_1, causing the disc tray to start moving out of the player. When the tray has moved about 3 mm, S_{904} opens, preventing power from being applied to the chucking motor M_2.

When the tray is fully open, or extended, disc-tray position switch S_{907} is closed. The input at pin 40 of IC_{102} is grounded (0 V). The output from pin 19 of IC_{102} is removed (0 V) and power is removed from loading motor M_1. This completes the unloading sequence.

7.3.2 Tray-Close Sequence

The following sequence of operation occurs from the disc-tray open state to the disc-tray closed state.

When open and close switch S_{903} is operated, the output at pin 20 of IC_{102} goes high. Because there is no inversion at pin 11 of IC_{304}, the output at pin 16 of IC_{304} goes high (about + 10 V).

The loading motor M_1 is supplied power through S_{905} (now in the loading position), and the disc tray is closed (moves within the player).

When the disc tray has reached the fully in position, S_{904} is closed, connecting M_2 to power through D_{964}. Chucking motor M_2 moves the chucking arm down over the disc. A short time after chucking is complete, S_{905} switches to the chucking position, shunting D_{964}.

Power to the loading motor M_1 is maintained throughout the chucking process

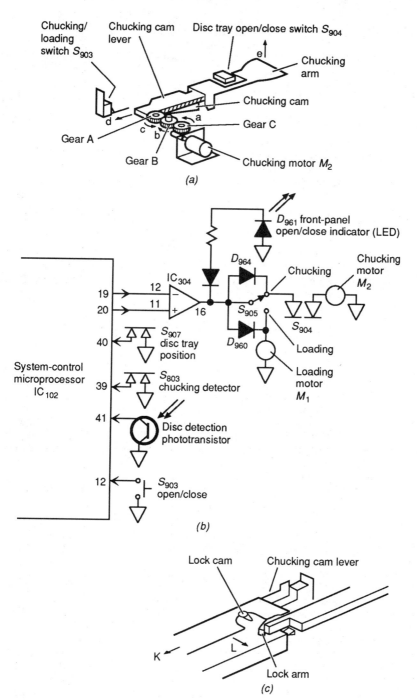

FIGURE 7.17 Horizontal front-load mechanical section (two load motors).

through D_{960}. This ensures that the disc tray does not move until locked (Sec. 7.3.4).

When the chucking arm reaches the final position (disc fully clamped on the turntable), chucking-detector switch S_{803} is closed. The input at pin 39 of IC_{102} is grounded (0 V). The output from pin 20 of IC_{102} is removed (0 V), and power (from pin 16 of IC_{304}) is removed from both motors. This completes the loading sequence.

7.3.3 Chucking-Arm Operation

Figure 7.17a shows the chucking mechanism at the moment when chucking is complete. The following sequence of operation takes place from the chucking complete state to the release of chucking state.

Chucking motor M_2 rotates, and gears A, B, and C rotate in directions a, b, and c.

The chucking-cam lever and the chucking cam move in direction d.

The chucking cam pushes the chucking arm in direction e. The chucking function is then released (disc unclamped).

When the chucking-cam lever reaches the final position, S_{905} switches over to the loading position, and the tray moves out.

7.3.4 Disc-Tray Fixing

During the initial phase of the chucking operation, the disc tray is fixed (or held) in position. This is shown in Fig. 7.17c. The sequence is as follows. At the beginning of the chucking operation, the chucking-cam lever moves in direction K. The lock arm is pushed in direction L by the chucking-cam lever, and the lock arm engages with the lock cam. The disc tray is then locked in the closed position.

7.3.5 Disc Detector

As shown in Fig. 7.17b, a disc-detection phototransistor is connected to pin 41 of IC_{102}. The purpose of the disc detector is to tell the system-control microprocessor whether or not a disc is in place.

If a disc is in place, the phototransistor is covered, and the input to pin 41 of IC_{102} is not grounded (high). If a disc is not in place, surrounding light is applied to the phototransistor, which conducts and grounds pin 41 of IC_{102} (producing a low at pin 41). These high and low signals at pin 41 tell IC_{102} the status of the disc.

In many players that have a disc detector, the turntable will not operate and the normal play function cannot be selected without a disc in place. Remember this when troubleshooting a "player will not play" symptom. Make sure that there is a disc in place and that light is not leaking to the disc-detector phototransistor (or that the phototransistor is not malfunctioning). Also remember that most players do not have a separate disc-detector system. Typically,

MECHANICS, ADJUSTMENT, AND REPLACEMENT   7.29

most present-day players will shut down when the laser cannot be focused (because there is no disc in place, as discussed in Sec. 5.6).

7.4 CDV MECHANICAL SECTION

This section is devoted to the mechanical functions of a typical CDV player (such as the player described in Chaps. 6 and 9). Again, these descriptions must be compared with the CDV player you are servicing (which, of course, will be altogether different).

Figure 7.18 shows major components of our CDV player (with covers removed), as well as a close-up view of the optical pickup assembly. The remainder of this section shows the steps necessary to replace and adjust those mechanical components most likely to require service (which, it is hoped, will never happen).

Caution: If it becomes necessary to work with the pickup assembly (especially if you remove and reinstall the pickup), always follow all of the precautions de-

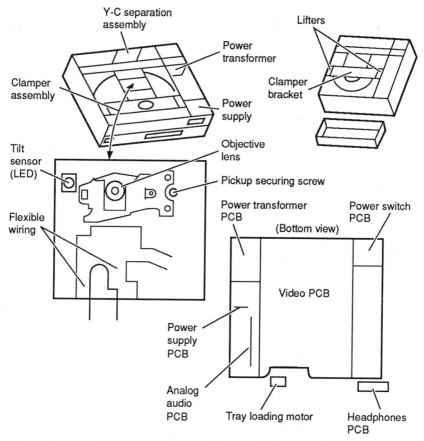

FIGURE 7.18 Major components of CDV player (with covers removed).

scribed in Sec. 4.1.5 and illustrated in Fig. 4.2. The laser-diode and photodiodes in any CDV optical pickup assembly are electrostatically sensitive devices (ESD) and must be so treated (using wrist straps, working on a conductive sheet, etc.).

7.4.1 Replacing the Pickup Assembly

This section describes the steps necessary to remove and replace the entire pickup assembly. Refer to Fig. 7.19 for the location of parts.

1. Remove the top cover and bottom plate by removing the screws, as shown in Fig. 7.19a.
2. Switch the power on and press the Open/Close key to eject the disc tray. Then switch the power off.
3. With the tray out, move the pickup assembly to the position shown in Fig. 7.19b. The pickup can be moved by rotating the slider motor manually. On most players, you can also operate the slider motor with a 1.5-V battery connected across the slider-motor terminals.
4. Stand the player on its side, as shown in Fig. 7.19c (with the power transformer at the top). Unfasten the three video-assembly screws and the three rear-panel screws. Open the video assembly and disconnect J_{402} from the digital-audio assembly.
5. Disengage the J_{101} lock in the servo assembly, and carefully remove the flexible cable. Place a paper clip across the foil conductors at the end of the flexible cable after removal from J_{101}.
6. Unfasten the pickup-securing screw from the player top, as shown in Fig. 7.19d. Carefully remove the pickup assembly. Try to avoid touching the soldered sections of the pickup assembly since these connect to the laser diode and photodiodes (which are ESD).
7. Using a wrist strap and conducting sheet, mount a new pickup assembly, and tighten the securing screw (Fig. 7.19d). Carefully reconnect (and lock) the flexible cable to J_{101} in the servo assembly (Fig. 7.19c).
8. If a new pickup assembly is installed, or even if the old pickup is returned to the player, check the adjustments described in Sec. 9.3. Pay particular attention to the spindle-motor centering check and adjustment in Sec. 9.3.6.

7.4.2 Dismantling the Tray Assembly

This section describes the steps necessary to remove and replace the tray assembly (which should be avoided). Refer to Fig. 7.20 for location of parts.

1. Remove the top cover (but not the bottom plate) by removing the screws, as shown in Fig. 7.20a.
2. Switch the power on and press the Open/Close key to eject the disc tray. Then switch the power off.
3. If the tray does not move to the fully out position (say because of a defective loading motor, no microprocessor open signals to the motor, etc.), use the manual tray-opening procedures in Sec. 7.4.3.

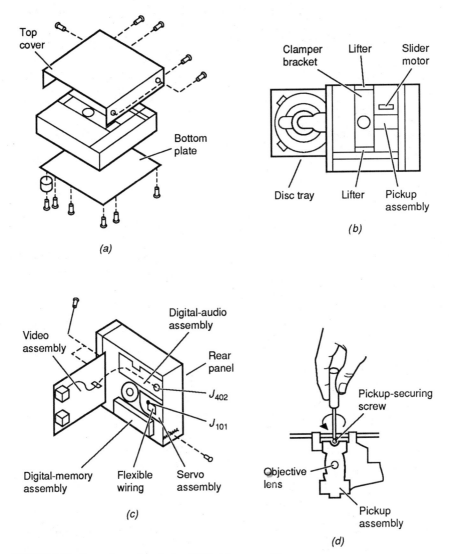

FIGURE 7.19 Removing and replacing CDV pickup assembly.

4. With the tray out, remove the screw, stopper, and bushing and then remove the tray assembly (by pushing the center of the tray slightly), as shown in Fig. 7.20b.

7.4.3 Manual Tray-Open Procedure

This section describes the steps necessary to open (or eject) the disc tray manually (when the tray cannot be moved out with the Open/Close key and loading motor).

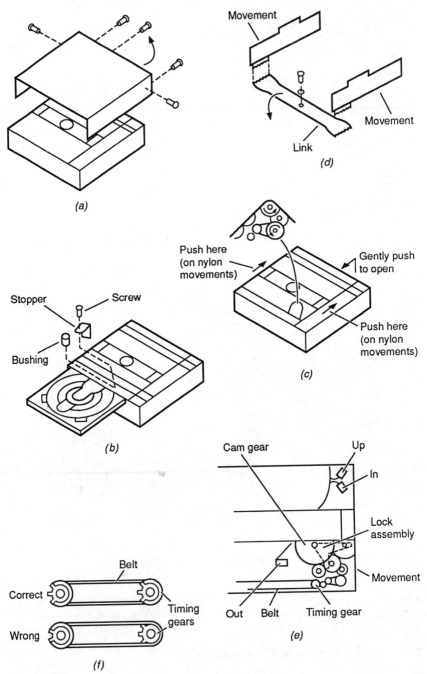

FIGURE 7.20 Removing and replacing CDV tray assembly.

1. With the top cover removed, push on the white nylon movements to raise the disc tray, as shown in Fig. 7.20c.
2. Turn either gear in the direction of the arrows until the tray just starts to move toward the front.
3. Push gently at the rear to move the tray to the full-out position.

7.4.4 Installing the Tray-Loading Mechanism

This section describes the steps necessary to install (or reinstall) the tray-loading components after the tray has been removed from the player.

1. Attach the link and left and right movements so that the three parts engage as shown in Fig. 7.20d. Then turn the link fully counterclockwise.
2. Engage the lock assembly with the rack-gear portion of the movement, and then attach the cam gear, as shown in Fig. 7.20e.
3. Attach the belt around the timing gears so that both timing gears face in the *same direction*, as shown in Fig. 7.20f.
4. Check that the Up and In switches are attached in the directions shown in Fig. 7.20e.

7.4.5 Attaching the Tray Assembly

This section describes the steps necessary to attach the tray assembly (if you should be so foolhardy as to remove it during service). Refer to Fig. 7.21 for the location of parts.

1. Switch the power on and press the Open/Close key to move the tray out. (Since there is no tray, watch the cam gear rotate). When the cam gear stops rotating, switch the power off while holding the Out sensor switch, as shown in Fig. 7.21a.
2. After making sure that the cam gear has rotated *fully clockwise* (as is the case when the tray is fully ejected, or out), check alignment of the loading-gear markers as follows: (a) the markers on the cam gear and tray-drive gear must be aligned and (b) the indentations (or markers) on the movements and lifters must also be aligned. Do not check either alignment until you are certain that the cam gear is fully clockwise. If the markers are not aligned, use the alignment procedure in Sec. 7.4.6.
3. With all gears properly aligned, open the front-panel door and insert the tray horizontally, aligning the tray with the timing gears on the left and right (Fig. 7.21a). When inserting the tray, make certain that the rollers are well seated in the grooves of the tray rack.
4. Continue pushing the tray until the first tooth of the tray rack engages the tray-drive gear teeth at the marker, as shown in Fig. 7.21b.
5. With the tray properly engaged, secure the stopper with the screw, and insert the bushing (Fig. 7.20b).

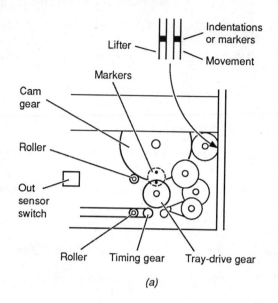

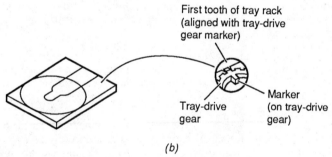

FIGURE 7.21 Attaching CDV tray assembly and aligning tray-loading gears.

6. Switch the power on and press the Open/Close key to move the tray in and out, making certain that the loading motor stops when the tray is fully in or out. There should be no problem if the In, Out, and Up switches are in the positions shown in Figs. 7.20e and 7.21a.

7.4.6 Aligning the Tray-Loading Gears

The tray-loading gears do not often go out of alignment during normal operation. Generally, you never need to check alignment unless the tray is removed from the player. However, if excessive force is applied to the tray, it is possible that

MECHANICS, ADJUSTMENT, AND REPLACEMENT 7.35

the gear teeth may slip (particularly if the force is applied when the teeth tips are engaged). Examples would be if the tray is moving out and hits against a solid object or if someone trys to push the tray in manually.

Should the gears go out of alignment, as evidenced by failure of the tray to go fully in or fully out (or both), check the alignment as shown in Fig. 7.21a. Then, if absolutely necessary, align the gears as follows.

If the markers of the cam gear and tray-drive gear are not aligned, remove the plastic washer that retains the tray-drive gear, and reset the gear as necessary for proper alignment.

If the markers of the lifter and movement are not aligned, rotate the cam gear clockwise (trying desperately not to damage the gear teeth) until just before the lock assembly has locked the tray-drive gear. Then align the markers (indentations) manually.

With both sets of markers aligned, repeat the tray-attaching procedures in Sec. 7.4.5 from the start.

CHAPTER 8
CD PLAYER TROUBLESHOOTING AND ADJUSTMENT

This chapter describes a series of troubleshooting and adjustment procedures for a cross section of CD player circuits. As discussed in the preface, it is not practical to provide specific troubleshooting procedures for every CD player. Instead, we describe a universal troubleshooting approach, using specific examples of CD players. These examples just happen to be the circuits discussed in Chap. 5. In this way, you can relate the theory (Chaps. 1 and 5) to the troubleshooting procedures in this chapter; then you can relate both to the specific CD player you are servicing.

Because adjustments are closely related to troubleshooting, we also describe typical adjustment procedures for CD players. Again, the circuits covered are some of those described in Chaps. 1 and 5 and we use the test equipment described in Chap. 4. When servicing other players, you must follow the manufacturer's troubleshooting and adjustment instructions exactly. Each type of player has its own electrical and mechanical test and adjustment points and procedures, which may or may not be different from procedures for other players.

Using the adjustment procedure example, you should be able to relate the procedures to a similar set of adjustment points on most CD players. Where it is not obvious, we also describe the purpose of the adjustment procedure. The waveforms or signals measured at various test points during adjustment are also included here. By studying the waveforms and signals, you should be able to identify typical signals found in most CD players, even through the signals may appear at different points for your particular player.

8.1 THE BASIC CD TROUBLESHOOTING FUNCTIONS

Troubleshooting can be considered as a step-by-step logical approach to locate and correct any fault in the operation of equipment. In the case of a CD player, seven steps are required.

First, you must study the player using service literature, user instructions, schematic diagrams, and so on, to find out how each circuit works when operating normally. In this way, you will know in detail how a given player should work. This is why the theory of operation for typical CD and CD-ROM players is included in Chap. 5.

Obviously, you must study the service literature for the particular player you are servicing. The functions and features of all CD players are similar, *but not identical*, to those of all other players. If you do not take the time to learn what is normal, you will never be able to distinguish what is abnormal. For example, some players simply produce better sound than other players, even when operating normally. (Frequency response, dynamic range, and signal-to-noise ratio are greater for one player.) You can waste hours of precious time (money) trying to make the inferior player perform like the quality instrument if you do not know what is "normal" operation. This is especially important when working on audio equipment, where all customers have a "golden ear."

Second, you must know the function of, and how to manipulate, *all player controls*. This is why the operating controls for typical CD players are discussed in Chap. 3. Again, you must learn the operating controls for the player being serviced. It is also assumed that you know how to operate the controls of the stereo system used to amplify and reproduce the CD layer output. An improperly adjusted stereo system can make a perfectly good player appear to be bad.

As an example, if the graphic-equalization controls of a stereo system are set to some weird combination, any CD player can sound equally weird. One suggestion for evaluation of a CD player *in the shop* is to have at least one stereo system of known quality. All players passing through the shop can be compared against the same standard. In any event, it is difficult, if not impossible, to check out a player without knowing how to set the controls. Besides, it makes a bad impression on the customer if you cannot find the disc tray, especially on the second service call.

Third, you must know how to interpret service literature and how to use test equipment. Along with good test equipment that you know how to use, well-written service literature is your best friend. In general, CD player service literature is good as far as procedures and drawings are concerned. Unfortunately, this literature is often weak when it comes to descriptions of how circuits operate (theory of operation). The "how it works" portion of most player literature is often sketchy, or simply omitted, on the assumption that you and everyone else know CD player theory as well as circuit functions.

Fourth, you must be able to apply a systematic, logical procedure to locate troubles. Of course, a "logical procedure" for one type of player is quite illogical for another. For example, it is quite illogical to check the loading-circuit microswitches for a top-load player (since such switches generally do not exist on top-load models). Likewise, many vertical-door players do not have laser safety interlock switches, and many players of all types do not have separate disc-detection circuits. However, all front-load players with horizontal trays have loading-circuit microswitches as well as laser safety interlocks. For this reason, we discuss logical troubleshooting approaches for various types of players, in addition to basic troubleshooting procedures.

Fifth, you must be able to logically analyze the information of an *improperly operating player*. For that reason, much of the troubleshooting information in this chapter is based on *trouble symptoms* and their relation to a particular circuit or group of circuits in the player, as discussed in Sec. 8.2. The information to be analyzed may be in the form of performance (such as failure of the disc to load normally) or may be indications taken from test equipment (such as waveforms or signals monitored with a scope). Either way, it is your analysis of the information that makes for logical, efficient troubleshooting.

Sixth, you must be able to perform complete checkout procedures on a player that requires service. Such checkout may be only a simple operation, such as selecting each mode of operation. At the other extreme, the checkout can involve complete adjustment of the player, both electrical and mechanical. This brings up a problem. Although adjustment of controls (both internal and front panel) can affect circuit operation, such adjustment can also lead to false conclusions during troubleshooting. There are two extremes taken by some technicians during adjustment.

On one hand, the technician may launch into a complete alignment procedure once the trouble is isolated to a circuit. No control, no matter how inaccessible, is left untouched. The technician reasons that it is easier to make adjustments than to replace parts. While such a procedure eliminates improper adjustment as a possible fault, the procedure can also create more problems than are repaired. Indiscriminate adjustment is the technician's version of "operator trouble."

At one extreme, a technician may replace part after part where a simple screwdriver adjustment will repair the problem. This usually means that the technician simply does not know how to perform the adjustment procedure or does not know what the control does in the circuit.

To take the middle ground, do not make any internal adjustments during the troubleshooting procedure until trouble has been isolated to a circuit and then make them only when the trouble symptom or test results indicate possible maladjustment. This middle-ground approach is taken throughout this chapter.

In any event, some checkout is required after any troubleshooting. One reason is that there may be more than one problem. For example, an aging part may cause high current to flow through a resistor, resulting in burnout of the resistor. Logical troubleshooting may lead you quickly to the burned-out resistor, and replacement of the resistor restores operation. However, only a thorough checkout can reveal the original high-current condition that caused the burnout. Another reason for after-service checkout is that the repair may have produced a condition that requires readjustment (such as after replacement of the pickup assembly or turntable motor).

Seventh, you must be able to use the proper tools to repair the trouble. As discussed in Chap. 4, CD player service requires all the common handtools and test equipment found in audio and stereo service, plus some special tools that are unique to the particular player. As a minimum, you must have (and be able to use) various metric tools, and you must have an assortment of test discs (at least some known-good discs). The average TV service technician is generally not familiar with these items (unless that technician also happens to service VCRs, videodisc players, tape recorders, stereo decks, etc.).

In summary, before starting any troubleshooting job, ask yourself these questions: Have I studied all available service literature to find out how the player works? Can I operate the player properly? Do I really understand the service literature and can I use all required test equipment and tools properly? Using the service literature and/or previous experience on similar players, can I plan out a logical troubleshooting procedure? Can I logically analyze the results of operating checks, as well as checkout procedures involving test equipment? Can I perform complete checkout procedures on the player, including electrical and mechanical adjustment and so on if necessary? Once I have found the trouble, can I use common handtools to make the repairs? If the answer is no to any of these questions, you simply are not ready to start troubleshooting any CD player. Start studying.

8.2 THE CD TROUBLESHOOTING APPROACH

The troubleshooting approach here is based on *trouble symptoms* and *grouping of circuits*. First, a separate section is devoted to each major circuit group in a typical CD player. A series of trouble symptoms is listed for each circuit group. The sections start with a description of the troubleshooting approach for that type of circuit and then go on to describe specific examples of trouble localization. Both the approach and symptoms can apply to any CD player but are related specifically to the player circuits described in Chaps. 1 and 5.

In some cases, the troubleshooting procedure requires adjustments, both electrical and mechanical. For that reason, the electrical adjustments are given ahead of the troubleshooting sections. (Mechanical adjustments, including the switches used in mechanical operation, are discussed in Chap. 7.) The adjustment procedures are referred to in the troubleshooting sections as necessary.

8.2.1 Preliminary Checks

Always make a few preliminary checks before launching into a full troubleshooting routine for any player. Start with the following:

Make certain that the transit screw is removed or loosened *before* operating the player. Tighten the transit screw only when the CD player is to be moved.

If practical, check that the customer's stereo system is operating normally before you do any extensive service on the CD player.

Cleaning the objective lens should be a routine part of servicing. A dirty objective lens can cause a variety of symptoms (intermittent or poor focus, skipping across the CD, erratic play, and excessive dropouts, to name a few). These same symptoms can also be caused by a defective disc. Try a known-good disc first.

Do not replace the pickup assembly or make any adjustments to the pickup before checking for mechanical problems that can affect it. For example, look for binding at any point on the pickup travel, which indicates that the rails, or guides, are adjusted too tightly (Sec. 8.4). At the other extreme, if you hear a mechanical "ratcheting" or "chattering" when the pickup is moved, the rails may be too loose. Note that mechanical adjustments are discussed in Chap. 7.

8.2.2 Preliminary Troubleshooting

Before you plunge into any CD circuit, here are some rather obvious but often overlooked checks that may cure mysterious problems:

If the CD player operates manually but not by remote control, check the remote-unit batteries (and control cable if any). Then try resetting the power circuits by pressing the front-panel power button on and off.

If the left or right speaker is dead when the CD player is used in a system, try playing another component connected to the audio system line (AM/FM tuner, cassette deck, etc.). If operation is normal with the other component, suspect the CD player. Also check for a loose cable between the CD player and other components. Then *temporarily reverse* the left and right speaker leads. If the same speaker remains dead, the speaker is at fault.

If the left or right speaker is dead when the CD player is used in a nonsystem configuration, temporarily reverse the left and right cable connectors at the amplifier input from the CD player. If the same speaker remains dead, the amplifier (or speaker) is at fault. If the other speaker goes dead, suspect the CD player or the player output cable.

If there is no sound from either speaker, check for the following. Both speaker-selector switches (if any) may be turned off. The wrong speaker terminals may have been selected (on amplifiers with two sets of speakers). The speaker cables may be disconnected. The CD player cables may be disconnected or improperly connected. In a nonsystem configuration, the output level may be too low (control too far counterclockwise), or the amplifier output selector may be set for the wrong source. Try playing the amplifier with a different input.

If the sound is distorted, the output level may be too high, or the CD player output may be connected to the phono input of the amplifier.

If there is hum or noise (only when the CD player is used), check for the following. A shield of the audio cable from the CD player may be broken, or the connector may not be firmly seated in the jacks. The CD player may be too close to the amplifier. The magnetic fields produced by the amplifier may induce hum into the player circuits (not likely, but possible). It is also possible (but not probable) that the laser may cause interference in the amplifier circuits.

If the player does not start, check the following (right after you are sure that the power cord is plugged in): Make certain that there is a CD in the tray, that the CD is properly loaded (not upside down; the label should be up), and the CD is firmly seated on the supports. Also check to see if the CD is very dirty, scratched, or warped. It is also possible that moisture has condensed on the CD or the objective lens.

If the sound cuts or repeats at some point, there may be a very dirty spot on the CD. Clean the CD with a soft cloth and mild detergent. (Do not use any commercial cleaners unless they are recommended for CDs.) It is also possible that there is a scratch on the CD. Try skipping the point where sound cuts or repeats.

8.3 CD ELECTRICAL ADJUSTMENTS

The following paragraphs describe complete adjustment procedures for a CD player such as the one described in Chap. 5. Each procedure is accompanied by diagrams that show the electrical locations for all adjustment controls and measurement points (test points, or TPs), as well as the waveforms or signals that should appear at the test points.

Remember that the procedures described here are the only procedures recommended by the manufacturer for that particular model of CD player. Other manufacturers may recommend more or less adjustment. It is your job to use the correct procedures for each player you are servicing.

Also remember that some disassembly and reassembly may be required to reach test and/or adjustment points. We do not include any disassembly or reassembly here for two reasons. First, such procedures are unique and can apply to only one model of player. More important, disassembly and reassembly (both electrical and mechanical) are areas where CD player service literature is generally well written and illustrated. Just make sure that you observe all the notes, cautions, and warnings found in the disassembly and reassembly instructions of the player literature. The procedures for removal of covers and gaining access to parts for some CD players are discussed in Chap. 7.

The remainder of this section describes the test and adjustment procedures for a typical CD player. Compare these procedures to those found in the service literature. We start with the laser-diode adjustment, which is always a good point to start on any type of CD player (including CD-ROM).

8.3.1 Laser-Diode Test and Adjustment

Normally, the laser diode need not be adjusted or tested unless (1) the pickup has been replaced or (2) troubleshooting indicates a laser problem. So, before you suspect the laser, consider the following points.

Even though the laser beam is invisible (except for the older helium-neon lasers), the diffused laser beam is often visible at the objective lens. (The lens appears to glow when the beam is on.) Also, when power is first applied to the optical circuits, the objective lens moves up and down two or three times to focus the beam on the CD, as described in Sec. 5.6. So, if you see the objective lens move when power is first applied, it is reasonable to assume that the laser is on and producing enough power to operate the optics.

Of course, this brings up some obvious problems. First, on most players, if you open the CD compartment and gain access to see the lens, you must override at least one interlock. Next, many players have some provision for shutting down the player optics if there is no CD in place (Sec. 5.6), so you must override this feature.

Most important, never, never look directly into the objective lens with power applied, and keep your eye at least 12 in from the lens. The purpose of the lens is to focus the beam sharply onto the CD. The lens can do the same job for your eye.

The service literature for early-model CD players sometimes recommends monitoring the laser with a light meter. However, it is more practical (and much easier) to adjust the laser diode output until you get an eight-to-fourteen modulation (EFM) signal of correct amplitude. This not only checks the laser but also checks the photodiodes and IC amplifiers following the photodidoes.

Figure 8.1a is the diagram for testing and adjusting the laser diode using the EFM signal. Before you make the adjustment, set R_{629} to minimum and then increase the setting as required.

Note that chuck switch S_3 must be in the closed (tray in) position before power is applied to Q_{601} and the laser. You must override S_3 manually during adjustment. If S_3 is in the tray-open position, the laser has no power and IC_{301} receives a 5-V signal to shut the system down.

1. Connect the scope as shown in Fig. 8.1a. With this connection you are monitoring the EFM signal (after the photodetector output is preamplified). As discussed, the EFM signal (at this test point) is also applied to the tracking, focus, and pickup-motor servos, as well as to the signal-processing circuits.

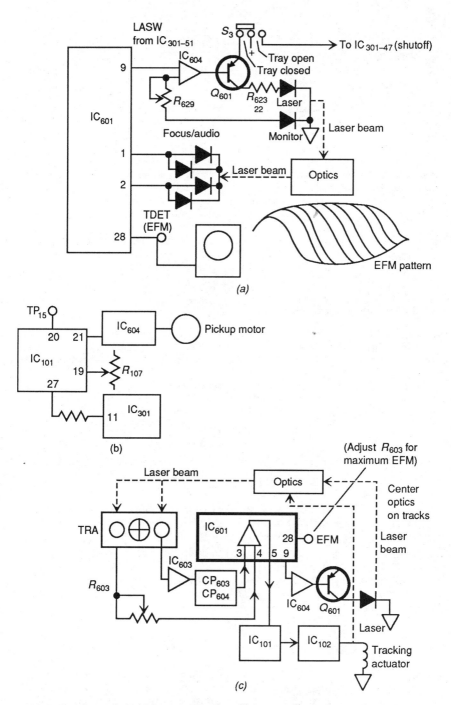

FIGURE 8.1 Laser-diode, pickup-motor, and tracking-servo adjustments.

2. Load a CD in the player and select play. The EFM signal should appear on the scope and produce a waveform similar to that shown in Fig. 8.1a.
3. Adjust R_{629} until the EFM signal level is 0.7 V (or as specified in the service literature, typically 0.5 to 0.9 V).

Be aware that laser diodes can be damaged by current surges (as can any semiconductor). Typically the lasers used in CD players have drive-current limits in the 40- to 70-mA range, possibly 100 mA. Generally, 150 mA is sufficient to damage (if not destroy) any CD laser diode.

Current limitations can present a problem since laser diodes may require more drive current to produce the required light as the diode ages. Some service literature spells out "safe" limits of laser drive current. The simplest way to check laser drive current is to measure the voltage across a resistor in series with the diode, such as R_{623} in Fig. 8.1a, and then calculate the drive current. For example, if the recommended laser diode current is 40 to 70 mA and the series resistance is 22 Ω, the voltage should be between 0.88 and 1.54 V. You can make this check before adjustment of the diode, and you should make the check after adjustment (to make sure that you have not exceeded the safe drive limits).

8.3.2 Pickup-Motor Adjustment

Pickup-motor adjustment is not available on all CD players. When available, the adjustment sets the point where the pickup accesses the beginning of the CD (the directory). If the adjustment is not correct, the program information may not be read properly.

Note that the adjustment controls the pickup-motor servo and is not to be confused with the inner-limit switch (S_1, Fig. 8.4). However, the two adjustments are interrelated. For example, if you set the switch so that the pickup motor cannot reach the inner limit, the servo cannot be adjusted to access the full CD directory.

1. Monitor the voltage at test point TP_{15}, as shown in Fig. 8.1b. With this connection, you are monitoring the pickup motor output from servo IC_{101}.
2. Load a CD in the player and select play mode.
3. While the CD is playing, connect pin 11 of IC_{301} to ground. This simulates a low TSW signal to IC_{301} from IC_{101} and prevents pin 11 from going high (which would cause IC_{301} to shut the system down).
4. Set the player to stop. After about 10 s, measure the dc level at TP_{15}, and adjust R_{107} so that the reading is 0 V ±50 mV. Adjust R_{107} in small increments and wait for the voltage level to stabilize before continuing the adjustment. (Make sure to remove the ground from pin 11 of IC_{301} when adjustment is complete.)

8.3.3 Tracking Adjustment

Figure 8.1c shows the tracking adjustment diagram. With this setup, you are monitoring the EFM signal and adjusting the optical pickup (through the servo and tracking actuator) so that the laser beam is properly centered on the tracks (as indicated when the EFM is maximum). Note that R_{603} sets the offset of the two tracking photodiodes but not the four remaining focus and audio photodiodes.

1. Load a CD in the player and select play mode. The EFM should produce a waveform on the scope (similar to that shown in Fig. 8.1a, the so-called "eye pattern").
2. Adjust R_{603} until the EFM is maximum (optical pickup centered on the tracks).

On some players, the display may become erratic and the audio may mute after this adjustment is made. If the display is erratic, set the player to stop and then go back to play. This should eliminate the erratic display.

Also note that the tracking adjustment should not be confused with the laser adjustment in Sec. 8.1.1. Adjustment of R_{629} (Fig. 8.1a) sets the level of the laser signal, while adjustment of R_{603} (Fig. 8.1c) positions the beam on the track for maximum signal. Some technicians simply adjust R_{629} for a maximum EFM signal (and possibly overdrive the laser in the process). This is the same as turning up the volume control on a radio that is not properly tuned in to a station.

8.3.4 Focus Adjustment

Figure 8.2a shows the focus adjustment diagram. With this setup, you are again monitoring the EFM, but you are now adjusting the optical pickup (through the servo and focus actuator) to properly focus the laser beam on the tracks (as indicated by maximum EFM). Note that R_{116} sets the offset of the four focus and audio photodiodes but not the two remaining tracking photodiodes.

1. Load a CD in the player and select play mode. The EFM should produce a waveform on the scope (similar to that shown in Fig. 8.1a).
2. Adjust R_{116} until the EFM is maximum (optical pickup focused on the CD tracks).

Again, if the display becomes erratic after this adjustment, stop and restart the player.

8.3.5 Turntable-Motor Adjustment

Figure 8.2b shows the turntable-motor adjustment diagram. With this setup, you are monitoring the drive signal to both coils A and B of the turntable motor (from motor-drive IC_{201}).

1. Load a CD in the player and select play mode.
2. Adjust R_{201} so that the output levels at DMCA and DMCB are equal. Usually DMCA and DMCB are about 2 V $_{p-p}$.

8.3.6 Dropout Sample and Hold Adjustment

Figure 8.2c is the adjustment diagram. This adjustment, not available on all players, is not to be confused with the sample and hold (S/H) audio circuits (Sec. 8.6).

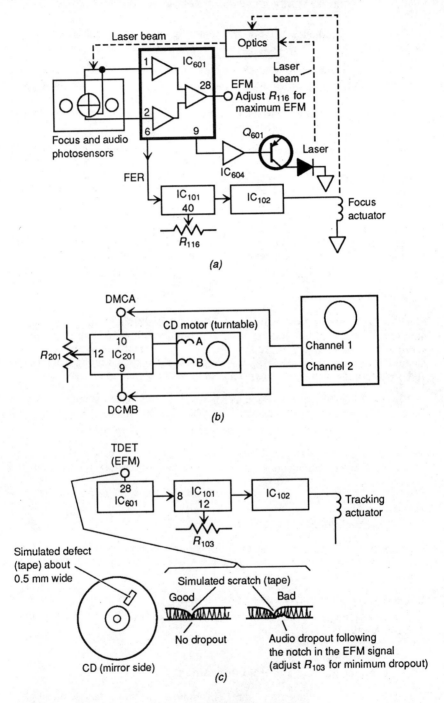

FIGURE 8.2 Focus-servo, turntable motor, and dropout S/H adjustments.

The S/H circuits shown in Fig. 8.2c are located in pickup servo IC_{101}, and they control the tracking-error signal (TER) signals (Sec. 8.8).

With the setup shown in Fig. 8.2c, play a CD with a simulated defect, and adjust the TER signals to produce the best response (minimum audio dropout). The effect is simulated by placing a black (nonreflective) tape on the mirror side of the CD. Then monitor the EFM and adjust for minimum dropout (ideally there should be no dropout).

You can make this adjustment by ear. The simulated defect produces a chattering or ticking in the audio. Adjust for *minimum noise*. The scope is generally more accurate (or you can monitor both ways). Do not turn up the volume with a simulated defect; the noise is unbearable.

1. Load a CD in the player and select play mode.

2. Adjust R_{103} for minimum audio dropout on the EFM display or for minimum chattering in the audio or both.

Note that with such a defect, a portion of the EFM display is cut out (typically a notch or wedge, starting from the top, as shown in Fig. 8.2c). However, you should be able to eliminate all (or most) of the audio dropout (as indicated by a cutout at the bottom of the EFM display). If you get considerable dropout at all R_{103} settings, IC_{101} may be defective.

8.4 MECHANICAL TROUBLESHOOTING APPROACH

Figures 8.3 and 8.4 show the components and circuits involved in the mechanical troubleshooting approach. In most players, the entire mechanism can be replaced as an assembly. Some manufacturers also recommend replacement of the motor and limit switches (and they describe the procedures for replacement and adjustment). The mechanical section is one area in which most CD player service literature is very good (if only the theory and troubleshooting sections were that clear).

We do not dwell on mechanical replacement and adjustment here. However, as a practical matter, *never disassemble the player mechanism beyond that point necessary to replace or adjust a part*. Similarly, never make any adjustments unless the troubleshooting procedures lead you to believe that adjustment is required.

If the tray refuses to open or close, first check that IC_{301} is getting the proper key-scan signals when the front-panel Open/Close button is pressed. If not, suspect the front-panel key matrix and/or IC_{901} (Fig. 5.1).

Next, make sure that the loading motor (LDM) receives a signal from pin 12 of IC_{102} when the Open/Close button is pressed. If it does, but the motor does not turn, suspect the motor (or possibly a jammed tray). If there is no signal at pin 12 when Open/Close is pressed, you have a problem between IC_{301} and the LDM.

Check for signals at pins 10 and 11 of IC_{102} each time Open/Close is pressed, and make sure that the signals invert (pin 10 high, then 11 low, then vice versa). Check for corresponding inverted signals at pins 33 (open) and 34 (close) of IC_{102}. If the signals are absent or do not invert when Open/Close is pressed, suspect IC_{301} (or possibly IC_{901} and the key matrix, Fig. 5.1).

If the tray opens but not fully, check when the open switch S_2 actuates, as

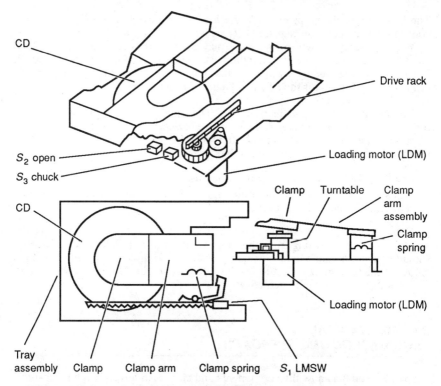

FIGURE 8.3 Major mechanical components of a CD player.

indicated by a low-to-high change at pin 48 of IC_{301}. If necessary, adjust S_2 (as described in the service literature and/or Chap. 7, Fig. 7.16). Before making any adjustments, check for a mechanical condition that might prevent the tray from opening fully (binding gears, jammed cross-rollers, improperly adjusted rails, etc.).

If the tray opens fully but the loading motor does not stop, the problem is usually an improperly adjusted S_2 (although there is an outside chance that IC_{301} is at fault).

If the tray closes but not fully and the clamp does not hold the CD in place on the turntable, check that S_3 actuates, as indicated by a high-to-low change at pin 47 of IC_{301}. If necessary, adjust S_3 as described in the service literature. Before adjustment, check for mechanical problems (something in the clamp hinges or tray wiring that has worked its way out of place, etc.).

If the tray closes and the clamp goes fully down but the LDM does not stop, the problem is probably an improperly adjusted S_3 but could be IC_{301}. Look for a high-to-low change at pin 47 of IC_{301}, which should occur when the tray is fully in and the clamp is down on the CD.

If the pickup does not move (with the tray in) when power is first applied (you may not be able to see the pickup, but you should *hear the motor*), check for slide/reverse (SLR) at pin 60 of IC_{301}. If SLR is absent, suspect IC_{301} or the lack of a reset signal at pin 24 of IC_{301} (from pin 40 of IC_{901}). If you get SLR but the motor does not run, suspect IC_{101}, IC_{604}, IC_{102}, and the motor itself.

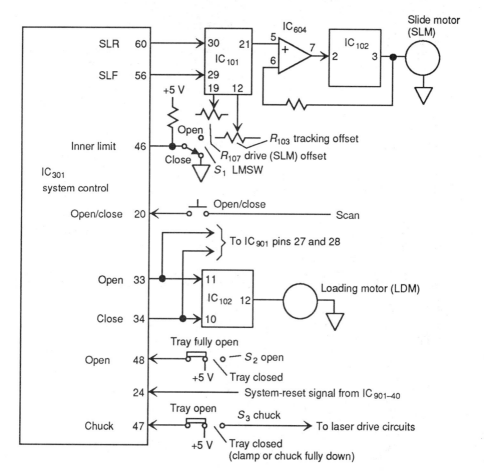

FIGURE 8.4 Circuits associated with mechanical components.

If the pickup appears to move to the inner limit when power is applied (you hear the motor stop and start) but the directory is not read properly (say that the total playing time or number of programs on the CD is not given on the front-panel display), try correcting the problem by adjustment of SLM offset R_{107} before going into the circuits. Follow the service-literature procedures for adjustment of R_{107} (or the procedures described in Sec. 8.1).

Note that on any problem with the pickup drive (forward or reverse) you should check for SLM drive voltage at pin 3 of IC_{102}. If the motor runs but the pickup does not move (or movement is erratic), look for mechanical problems (jammed gears, binding rollers, improperly adjusted rails, etc.).

If the pickup moves but does not reach the inner limit, check when S_1 actuates, as indicated by a high-to-low change at pin 46 of IC_{301}. If necessary, adjust S_1 as described in the service literature and/or Chap. 7.

Before adjusting S_1, check the adjustment of the SLM offset R_{107}. If R_{107} can be adjusted so that the pickup accesses the CD properly (the directory can be read in full), the S_1 adjustment is probably good. On most players, S_1 does not go out of adjustment except when the pickup is replaced (or when there has been tampering).

If the pickup reaches the inner limit but the motor does not stop, the problem is almost always one of an improperly adjusted S_1. A possible exception is where IC_{301} is defective and is not responding to the low at pin 46.

8.5 LASER TROUBLESHOOTING APPROACH

Figure 8.5 shows the components and circuits involved in the laser troubleshooting approach. The laser diode must produce a proper beam if the CD player is to perform all functions correctly. If the beam is absent, there is no EFM signal. If the beam is weak, EFM is weak. If the monitor diode does not monitor the laser properly, the beam can shift to an incorrect level (high or low) without being sensed by the laser-drive circuits. Any of these conditions can cause improper tracking which, in turn, can produce an even weaker EFM.

Always look at the laser circuits first when you have mysterious symptoms with no apparent cause (improper tracking that cannot be corrected by adjustment, excessive audio dropout with a known-good CD, etc.). Start with laser-diode adjustment. This should show any obvious problems in the laser circuits and also tell if the EFM signal is good. (An EFM signal of proper amplitude generally means a good laser.)

If the laser appears to be completely dead (no glow at the objective lens, no EFM, or no movement of the focus when power is first applied, as described in Sec. 8.7), make sure that Q_{601} is getting 5 V through S_3. If not, suspect S_3 (or adjustment of S_3). Note that when the tray is open, pin 47 of IC_{301} goes high (Fig. 8.4), shutting down many functions of IC_{301}, including the laser switch (LASW) signal at pin 51 (Fig. 8.5).

If power is applied to the laser diode, look for an LASW signal at pin 51 of IC_{301}. If LASW is high (absent), suspect IC_{301}. If it is present, check for a signal at pin 3 of IC_{604} from the monitor diode. If it is abnormal, suspect the monitor diode and/or R_{629}.

If the signals are present at both pins 2 and 3 of IC_{604}, look for drive signals at pin 1 of IC_{604} and at the base of Q_{601}. If they are absent, suspect IC_{604}. If they are present but there is no laser-beam output, suspect Q_{601} or the laser diode. An alternate method for checking the laser diodes is described in Sec. 8.3.1.

Finally, check for EFM signals at pins 28, 23, and 20 of IC_{601} and EFM square wave (EFMS) signals at pin 15 of IC_{601}. (Note that the EFM signals are of high frequency, while the EFMS signals are square waves).

If the EFM signals are absent at pins 23 and 28 of IC_{601}, suspect IC_{601} or the photosensors. If the signals are present at pins 23 and 28 but not at pin 20, suspect IC_{602}.

If the signals are present at pin 20 of IC_{601} but not at pin 15, suspect IC_{601}. However, make sure that the data-slice level control (DSLC) and preference pulse (PREF) signals (both square waves) from IC_{402} are present at pins 21 and 22 of IC_{601}. If DSLC and PREF are absent, suspect IC_{402}. (Unfortunately, IC_{402} does not generate PREF and DSLC signals if there is no EFMS applied to pin 47 of IC_{402}.)

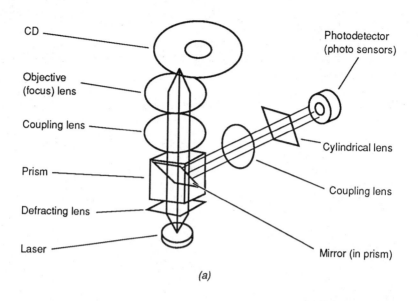

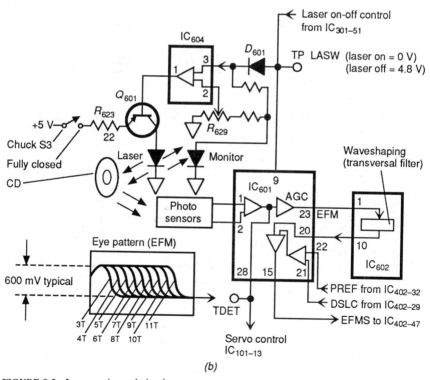

FIGURE 8.5 Laser optics and circuits.

8.6 SIGNAL-PROCESS TROUBLESHOOTING APPROACH

Figure 8.6 shows the components and circuits involved in signal-processing troubleshooting. A failure in the signal-process circuits can cause a variety of failure symptoms in both the audio and turntable-motor circuits. Similarly, a failure in the system control can appear as a failure in the signal processing. From a practical standpoint, there is no sure way to tell if the problem is in the signal processing, system control, turntable motor, or audio. However, there are some checks that can help you pin down the problem.

First check for audio at pin 17 of digital-to-analog (D/A) converter IC_{403}. You should get both left- and right-channel audio (at a very low level). If you get no measureable audio, the problem is likely to be in the audio circuits (Sec. 8.7).

Next, if there are excessive audio dropouts (with a known-good CD) and the front-panel indications are not normal (such as the time-code changing as the CD continues to rotate), the problem is probably in the signal-process circuits. Check all of the waveforms to and from the signal-process circuits shown in the service literature. Pay particular attention to the following (using Fig. 8.6 as a guide):

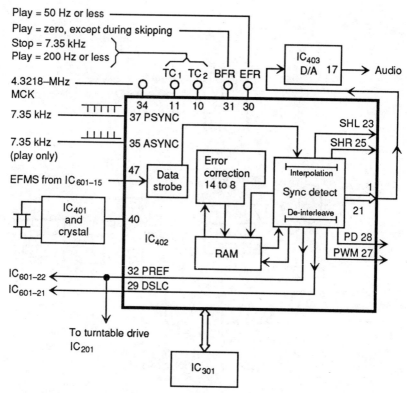

FIGURE 8.6 Laser signal-processing circuits.

Check for a 4.3128-MHz master clock (MCK) signal at pin 34 of IC_{402}. If it is missing, check the crystal and IC_{401} connected at pin 40 of IC_{402}.

Check for 7.35-kHz Psync and Async signals at pins 37 and 35 of IC_{402}. The Async signal should be present only during play, but Psync should be available in both stop and play.

Make certain that PREF and DSLC are supplied to IC_{601} (Sec. 8.5) and returned to IC_{402} as square-wave EFMS signals. If EFMS is missing, check for high-frequency EFM signals at pin 20 of IC_{601}, as described in Sec. 8.5.

Check all signals between IC_{301} and IC_{402}. It is not practical to analyze the waveforms of these signals. However, if you can measure a data stream (pulses) on each line with a scope, it is reasonable to assume that the signal is correct. If one or more of these signals are missing, suspect IC_{301}, IC_{402}, or both. Remember that a signal from IC_{301} can depend on a signal from IC_{402} and vice versa. So you may have to replace both ICs to find the problem. Also remember that IC_{301} may not produce the signals unless other signals (such as focus OK, FOK, and tracking OK, TOK) are applied to IC_{301}, as discussed in Secs. 8.8 and 8.9.

Before you pull IC_{402}, check the TC_1 and TC_2 signals at pins 11 and 10 of IC_{402}. Both of these test points (which indicate the accuracy of the decoding process within IC_{402}) should produce a 7.35-kHz signal during stop but then drop to 200 Hz or less when play is selected. If not, suspect IC_{402}.

Finally, check the block error (BFR) and EFR signals at pins 31 and 30 of IC_{402}. Both of these test points show the accuracy of the sync and detection functions within IC_{402}. In play, BFR should always show zero, except under conditions of excessive skipping across the CD. In play, EFR may produce a signal but at a frequency below 50 Hz. The EFR and BFR signals are not defined during stop.

8.7 AUDIO TROUBLESHOOTING APPROACH

Figure 8.7 shows the components and circuits involved in audio troubleshooting. *If there is no audio at any output, including the headphone jack*, start by checking for audio at pin 17 of IC_{403}. If it is absent, suspect the signal-process circuits (Sec. 8.6).

Next, check the SHR and SHL signals from IC_{402} (pins 23 and 25). If SHR and SHL are present and there is audio at pin 17 of IC_{403}, trace the audio signal level from IC_{403} to the headphones and/or rear-panel jacks. Note that the level for audio at both the rear-panel jacks and headphones is controlled by output control R_{524}.

Note that if relays RY_{501} and RY_{903} are not operating properly, the audio is cut off from the rear-panel jacks (but not the headphones). However, if the switches in IC_{506} are not responding properly to emphasis signals from pin 41 of IC_{402} or if the signal is missing, the audio passes but may appear distorted. So if you get a "the audio appears distorted when I play certain CDs" trouble symptom, start by checking the emphasis newtork and IC_{506} as well as the emphasis signal (EMP) from pin 41 of IC_{402}.

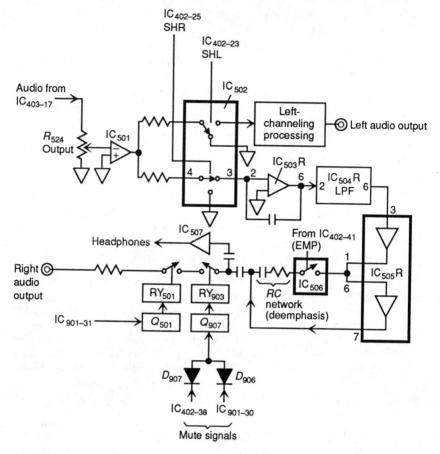

FIGURE 8.7 Audio circuits.

8.8 AUTOFOCUS TROUBLESHOOTING APPROACH

Figure 8.8 shows the components and circuits involved in autofocus troubleshooting. If you suspect problems in the autofocus system, put in a CD, press Play, and check that the pickup moves up and down two or three times and then settles down. If it does not, check that the laser is on, as discussed in Sec. 8.5. If the laser is on but there is no focus, try adjusting the focus servo, as described in the service literature and/or Sec. 8.3.4.

Next, check the focus actuator by measuring resistance of the focus and tracking coils, as described in Sec. 7.1.1 (Fig. 7.2). If the actuator coils appear to be good and the problems cannot be corrected by adjustment, check the autofocus circuits as follows:

If the focus actuator does not move up and down, check for focus up-down (FUD) pulses just after play is selected. If the FUD pulses are absent at pin 50

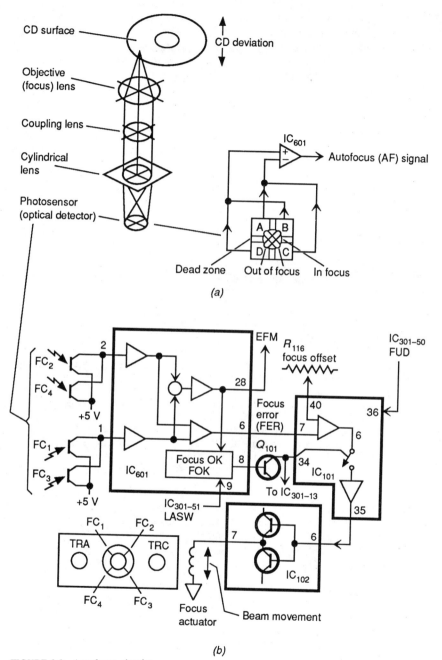

FIGURE 8.8 Autofocus circuits.

of IC_{301} and/or pin 36 of IC_{101}, suspect IC_{301}. Next, check for pulses at pin 35 of IC_{101}, pins 6 and 7 of IC_{102}, and at the focus coil.

If the focus actuator moves but focus is not obtained, check for FOK signals at pin 34 of IC_{101}, pin 13 of IC_{301}, and pin 8 of IC_{601}. (If the FOK signals are not present, IC_{301} should shut the system down.)

If the FOK signals are absent, suspect IC_{601} or possibly the four pickup sensors FC_1 through FC_4. If FOK signals are absent at $IC_{101\text{-}34}$, suspect Q_{101}. Also note that the FOK signal is not generated unless there is an LASW signal applied to pin 9 of IC_{601} from IC_{301}.

Next, check for FER signals at $IC_{601\text{-}6}$. If FER signals are absent at $IC_{601\text{-}6}$, suspect IC_{601} or possibly the sensors FC_1 through FC_4. (However, if FOK signals are present, FC_1 and FC_4 are probably good.)

If you suspect FC_1 and FC_4, monitor the EFM signals at pin 28 of IC_{601}. If the EFM is good, it is reasonable to assume that all four sensors FC_1 through FC_4 are good.

8.9 LASER-TRACKING TROUBLESHOOTING APPROACH

Figure 8.9 shows the components and circuits involved in laser-tracking troubleshooting. It is often very difficult to separate tracking and focus servo problems. For example, unless there is an FOK signal applied to $IC_{101\text{-}34}$ through Q_{101}, the TER signal does not pass to the tracking actuator. Both the focus and tracking servos use the laser beam as a source of error signal, although different sensors are used. To complicate the problem further, the TER signal is also used by the slide motor (SLM) as a fine speed control. (This is done in IC_{101}.) If TER is lost, both the tracking actuator and SLM have no control signals. Either condition can produce symptoms of improper tracking.

The TER signal is passed through variable-gain and error-detection circuits in IC_{101}. These circuits can interrupt the TER signal when errors are detected (in the CD or because of improper tracking). Failure in any of these circuits can cut off, or alter, the TER signal, making it appear that either tracking or SLM servos are at fault (when actually the servos are good).

First try correcting any tracking problems with adjustment of R_{603} and R_{103}, as described in the service literature or in Sec. 8.3. Then make a quick check of the tracking-actuator coil, as described for the focus-actuator coil in Sec. 8.8. Finally, see if the SLM moves to the inner limit when power is first applied, as described in Sec. 8.4 (possibly adjust R_{107} and S_1). If the SLM moves to the inner limit, this confirms that the SLM, reset circuit, and basic servo circuits of IC_{101} and IC_{102} are good.

If the SLM is operating and the tracking-actuator coil appears to be good but tracking problems cannot be restored by adjustment, check the tracking and pickup (SLM) motor.

Check the TER signal from the source to the tracking-actuator coil (and drive signal to the SLM). Pin 5 of IC_{601} is a good place to start. Then check the tracking-actuator coil.

If TER is present at pin 5 of IC_{601} but not at the coil, suspect IC_{101} and IC_{102}. Look for a drive signal at pin 23 of IC_{101}. If it is present, the problem is probably in IC_{102}. If it is absent, the problem is localized to IC_{101}.

Check for drive signals at the SLM. If TER is present at the tracking coil but

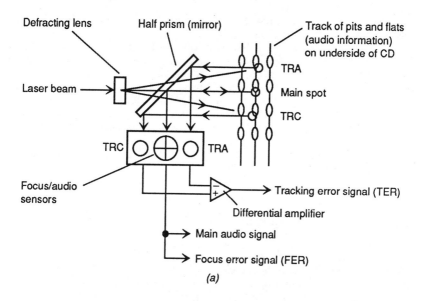
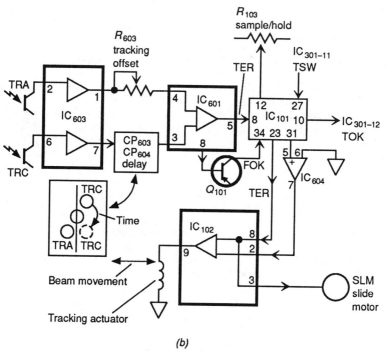

FIGURE 8.9 Laser tracking circuits.

there is no drive at the SLM motor, suspect IC_{102} and IC_{604} (Sec. 8.4). Also check for drive signals at pin 21 of IC_{101}.

Before you pull IC_{101}, remember that IC_{101} must receive a number of signals before TER signals can pass. For example, IC_{101} must be turned on by FOK and TSW signals, and IC_{301} must receive TOK signals from IC_{101} before returning the TSW signals. Also, the TER signals are analyzed by error-detection circuits within IC_{101}. If any of these signals or voltages are absent or abnormal, IC_{101} remains cut off, and the TER signals do not pass. So always check the signals and voltages at the pins of IC_{101} (using service-literature values) before you decide that IC_{101} is defective.

If TER is absent at $IC_{601\text{-}5}$, suspect IC_{601}, CP_{603}, CP_{604}, IC_{603}, and the TRA and TRC sensors.

8.10 TURNTABLE TROUBLESHOOTING APPROACH

Figure 8.10 shows the components and circuits involved in turntable troubleshooting. It is not difficult to tell if the turntable motor fails to rotate. Similarly, the cause of such a total failure is generally simple to locate. For example, you can check DMCA and DMCB at pins 9 and 10 of IC_{201} for drive signals to the motor windings. If the drive signals are present but the motor does not turn, suspect the motor. If either of the drive signals is not present, trace from the motor to IC_{201} and from IC_{402} to IC_{201}.

Before you decide there is a problem in the turntable motor circuits, remember that DMSW, CLVH, and ROT signals must come from IC_{301}. Also, if IC_{301} does not receive an FOK signal from the autofocus circuits (Sec. 8.8), the DMSW, CLVH, and ROT signals are set to prevent IC_{402} and IC_{201} from passing the PREF, PWM, and PD signals to the motor circuits.

Both DMSW and CLVH are made low to turn on the motor when play is selected. ROT goes low for about 1 s after play is selected. If DMSW, CLVH, and ROT all remain high after play is selected, check for FOK signals to IC_{301} (at pin 13) and TOK signals to $IC_{301\text{-}12}$. Of course, if only one of the three signals remains high, IC_{301} is most likely at fault.

The problem is not quite that simple if the motor rotates, but you are not sure of the correct speed. This is especially true when you consider that the motor speed is constantly changing. You must rely on waveform measurements and adjustments. So the first logical step in motor troubleshooting is to perform adjustment of R_{201}, as described in the service literature or Sec. 8.3.1.

If you get the DMCA and DMCB drive signals and the motor is turning (indicating that the DMSW, ROT, and CLVH from IC_{301} are good) but you are unable to set the output levels as described, check all waveforms associated with the motor-control circuits as follows.

Check the PWM, PREF, and PD signals from IC_{402}. If any of these are absent or abnormal, suspect IC_{402}. Next, trace signals between IC_{201} and the motor (Hall-effect and drive signals). If any of these signals are absent or abnormal, suspect either IC_{201} or the motor.

Note that PREF from IC_{402} is also applied to IC_{601}, along with the DSLC signal from IC_{402}, to form the EFMS signal that is returned to IC_{402} (Sec. 8.5). If the EFMS signal is absent, IC_{402} does not produce PREF, PWM, and PD signals. Of

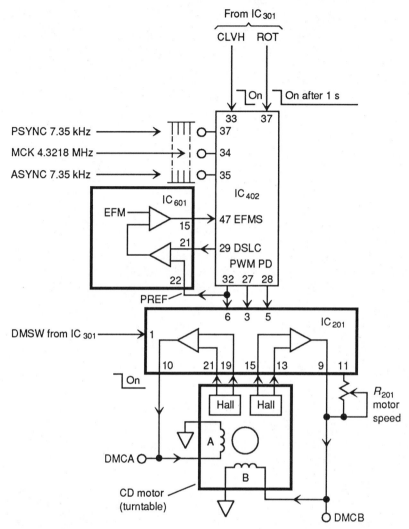

FIGURE 8.10 CD turntable motor circuits.

course, if the EFMS signal is not applied to IC_{402}, several other problems occur (no audio, etc.).

One way to check if the EFMS signal is being processed properly is to compare the Psync and Async signals at pins 37 and 35 of IC_{402}, using a dual-trace scope. As shown in Fig. 8.10, both signals should be synchronized as to time. If they are not, or if either signal is missing, suspect IC_{402} (or possibly IC_{601}).

As you can see, the turntable motor circuits are closely interrelated with the signal-processing IC_{402} circuits (Sec. 8.6). A failure in signal processing can also cause the motor circuits to appear defective. So if you are unable to locate a problem in the motor circuits, try checking the signal-processing circuits.

CHAPTER 9
CDV PLAYER TROUBLESHOOTING AND ADJUSTMENT

This chapter describes a series of troubleshooting and service notes for a cross section of CDV players. As discussed in the preface, it is not practical to provide a specific troubleshooting procedure for every CDV player. Instead, we describe a universal troubleshooting approach, using specific examples of CDV players. These examples just happen to be the player discussed in Chap. 6. In this way, you can relate the theory of Chap. 6 to the troubleshooting procedures in this chapter; then you can relate both to the specific CDV player that you are servicing.

Because adjustments are closely related to troubleshooting, we also describe typical adjustment procedures for CDV players. Again, the player covered is that described in Chap. 6, using the test equipment and tools described in Chap. 4. When servicing other players, you must follow manufacturer's service instructions exactly. Each type of player has its own electrical and mechanical adjustment points and procedures, which may or may not be different from procedures for other players.

Using the adjustment procedure examples, you should be able to relate the procedures to a similar set of adjustment points on most CDV players. Where it is not obvious, we also describe the purpose of the adjustment procedures. The waveforms measured at various test points during adjustment are also included here. By studying these waveforms, you should be able to identify typical signals found in most players, even though the signals may appear at different points for your particular unit.

9.1 THE BASIC CDV TROUBLESHOOTING FUNCTIONS

The basic CDV troubleshooting functions are essentially the same as for CD players (as described in Sec. 8.1) and are not repeated here. However, remember that a CDV player produces *both audio and video* which must be monitored during service. It is therefore essential that you have both a shop-standard stereo system and a shop-standard TV or monitor of known quality. All players passing through the shop can be compared against the same standard. The monitor should be capable of displaying S-VHS (with separate Y and C signals). Many late-model CDV players (including the player described in Chap. 6) can produce both conventional TV signals (composite video and/or RF) as well as S-Video with separate Y and C.

One practical way to confirm troubles in a CDV player is to monitor a conventional TV broadcast with a TV set or monitor and compare the playback quality with that of the broadcast. However, assuming a good test disc, the player picture is usually better than a broadcast picture (and considerably better than a VCR picture). Also, remember that an improperly adjusted monitor or TV can make a perfectly good player appear to be bad. This is a major problem in servicing CDVs (and VCRs) and is another good reason for a shop-standard monitor or TV.

9.2 THE CDV TROUBLESHOOTING APPROACH

The troubleshooting approach for our CDV player is based on functional circuit groups. The CDV players of all manufacturers have certain circuit groups in common (power supply, spindle drive, system control, servo, video/TBC, memory video, audio, and mechanical). In this chapter, a separate section is devoted to each of the major circuit groups. These groups are essentially the same as the circuit groups discussed in Chap. 6. Often, direct reference is made to the diagrams in Chap. 6.

Using this section and circuit-group approach, you can quickly locate information needed to troubleshoot a malfunctioning CDV player. For example, in Chap. 6, you find (1) an introduction that describes the purpose or function of the circuit and (2) some typical circuit descriptions or circuit theory. In this chapter, you find a logical troubleshooting approach for the circuit (based on manufacturers recommendations).

In many cases, the troubleshooting procedure requires adjustment, both electrical and mechanical. For that reason, the adjustment procedures for our CDV player are given in Sec. 9.3. These adjustments are referred to in the troubleshooting procedures as necessary. In addition, Sec. 9.4 summarizes trouble symptoms that can be caused by improper adjustment.

9.3 CDV ELECTRICAL ADJUSTMENTS

The following paragraphs describe complete adjustment procedures for a CDV player like the one described in Chap. 6. Each procedure is accompanied by diagrams that show the electrical locations for all adjustment controls and measurement points (test points), as well as the waveforms or signals that should appear at the test points.

Remember that the procedures described here are the only procedures recommended by the manufacturer for that particular model of CDV player. Other manufacturers may recommend more or less adjustment. It is your job to use the correct procedures for each player you are servicing.

Also remember that some disassembly and reassembly may be required to reach test and/or adjustment points. We do not include full disassembly and reassembly here for two reasons. First, such procedures are unique and can apply to only one model of player. More important, disassembly and reassembly (both electrical and mechanical) are areas in which CDV player service literature is generally well written and illustrated. Just make sure that you observe all of the notes, cautions, and warnings found in the disassembly and reassembly sections of the player service literature. The procedures for removal of covers and gaining access to parts for our CDV player are discussed in Sec. 7.4.

9.3.1 Fixtures and Tools Required

The fixtures and tools required for each test are described as part of the following test procedures.

9.3.2 Preparations and Precautions for Adjustment

Carefully observe all of the following precautions *before* attempting any adjustment:

Player settings: For most adjustment procedures the player should be placed on a service test stand (Sec. 7.4) or stood on its side with the power transformer at the top. Both the video and digital audio assemblies should be open.

Double clamping: If double clamping (Sec. 6.6) is performed while the player is standing on its side, the disc will slip out of position. This can be prevented by placing the player in the test mode, as described in Sec. 9.3.3.

Test discs: The CDV or LD test discs used in these adjustments may be either Pioneer N series or F series. The frame numbers given in the text are N series numbers, while those enclosed in parentheses are F series numbers.

Tracking and slider servo on-off: The tracking and slider servos can be opened (tracking off) or closed (tracking on) using the test mode (Sec. 9.3.3) without disconnecting the servo loops.

Oscilloscope settings: Unless otherwise specified, all scope settings shown in the adjustment diagrams are values obtained with a 10:1 probe.

RF interference: When the bottom (video assembly) circuit board is open (Sec. 7.4), a diagonal bar (interference) will appear on the display when the RF modulator or converter is used. This condition is normal because the RF modulator is not fully grounded to the main chassis. Adding a separate ground wire will help but may not eliminate the interference when operating the player disassembled.

9.3.3 Test Mode Operation

The player may be placed in a test mode using the remote transmitter keys (Fig. 3.4). The following is a summary of the test mode functions and displays.

Entering the Test Mode. To enter the test mode, press and hold the favorite track selection (FTS), Clear, and Store keys, and then press Power On.

Test Mode Operations. The following operations occur in the test mode:

Player operation is normal.

When the disc is clamped, the clamping operation is not repeated twice (called double clamping).

When the Eject key is pressed in test mode, the disc stops but is not ejected. (With CD and CDV Single discs the tray can be ejected by pressing the Eject key twice.)

When the Emphasis signal (CX) on a CD, CDV Single, or CD with digital sound is detected in the test mode, an asterisk (*) is displayed at the top right of the display.

When the player is in a test mode, the remote transmitter keys can be used to provide various operations. The following is a summary of such operations:

The front-panel fluorescent display can be toggled between a normal display and an all-lamps-on display using the 0 key.

The focusing operation (such as described in Secs. 6.6 and 6.8) can be switched on and off by the 1 key. Each press of the 1 key alternately toggles focus on and off. If focusing cannot be obtained after five trials, the focusing operation stops.

The slider motor can be moved in or out by the 2 key. Each press of the 2 key alternately activates or deactivates the manual-operation mode. With manual operation activated, the slider can be moved in or out by pressing the FF/Rev key. Unless the slider motor is in the manual-operation mode, the FF/Rev key operations in the test mode are the same as in normal (nontest) modes.

The slider motor and tracking servos can be switched on and off, after the play mode is accessed, by the 3 key.

The tilt servo can be switched on and off by the 4 key.

The memory-through mode (Sec. 6.12.6) can be switched on and off by the 5 key.

Internal information in the main microcomputer (QU_{01}, Fig. 6.7) can be displayed on the fluorescent display by pressing the 6 key. The display format is shown in Fig. 9.1.

The 7 key is not always used in the test mode. However, on some early models of our player, certain software credits are displayed on the monitor screen when the 7 key is pressed. Should this occur, exit the selected test mode by pressing the power off. If the player is in the vertical position when power is turned off, the tray may double clamp, allowing the disc to fall out; be careful to hold onto the disc.

The FTS RAM (QU_{08}, Fig. 6.7) can be checked using the 8 key. All areas of the FTS RAM are checked (write and read). If the result is good, the FTS display blinks for 5 s. If not, the FTS display does not come on. Note that the FTS program data is not destroyed when the RAM is tested by the 8 key. The exiting data bits are written back into the RAM simultaneously. (If they are not, you have a real problem.)

The test mode is terminated when the 9 key is pressed. This is not a toggle function. Once the 9 key is pressed, the test mode cannot be resumed until the power is switched off and the test mode is again entered (pressing and holding the FTS, Clear, and Store keys and then pressing Power On).

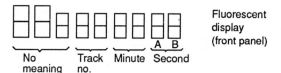

```
  No        Track    Minute  Second
meaning     no.                A  B
```

Track No.: The current key code or remote-control code number received by QU_{01} displayed in two digits

Minute: The selected internal operation mode (test mode) of QU_{01} displayed in two digits:

    00 = Park mode   02 = Open mode   04 = Setup mode
    06 = Play mode   08 = Scan mode   10 = Search mode

Second: Sections A and B represent the control status as follows:

Section A

Control Display	CDV/CD	Tilt servo	Scan	Focus
0	CD	Off	Rev	Off
1	CDV	Off	Rev	Off
2	CD	On	Rev	Off
3	CDV	On	Rev	Off
4	CD	Off	Fwd	Off
5	CDV	Off	Fwd	Off
6	CD	On	Fwd	Off
7	CDV	On	Fwd	Off
8	CD	Off	Rev	On
9	CDV	Off	Rev	On
–	CD	On	Rev	On
E	CDV	On	Rev	On
n	CD	Off	Fwd	On
d	CDV	Off	Fwd	On
"	CD	On	Fwd	On
(blank)	CDV	On	Fwd	On

Section B

Control Display			Tracking	T-count divider
0			Close	1/1
1			Close	1/1
2			Close	1/1
3			Close	1/1
4			Open	1/1
5			Open	1/1
6			Open	1/1
7			Open	1/1
8			Close	1/256
9			Close	1/256
–			Close	1/256
E			Close	1/256
n			Open	1/256
d			Open	1/256
"			Open	1/256
(blank)			Open	1/256

FIGURE 9.1 Display format for internal information in main microcomputer QU_{01}.

9.6　　　　　　　　　　　　　　　　CHAPTER NINE

Test Mode Sequence. Many of the following test and adjustment procedures require various operations of the test mode. Also, many of the procedures must be performed in a certain sequence. The following is a summary of that sequence. If the rough-grating (Sec. 9.3.4), spindle-motor centering (Sec. 9.3.7), pickup tracking-direction (Sec. 9.3.8), and fine-grating (Sec. 9.3.12) adjustments are performed, all mechanical or servo adjustments (Sec. 9.3.4 through 9.3.16) must be performed in sequence.

9.3.4 Rough-Grating and Tracking-Balance Adjustments

Figure 9.2 shows the rough-grating and tracking-balance adjustment diagram. Figure 6.18 shows additional circuit details. This adjustment sets the laser beam to the optimum position on the disc tracks and sets the tracking-servo offset voltage to 0 V.

Rough-Grating Adjustment. Access the test mode and play a CDV (LD) test disc. Operate the keys to display the frame number on the monitor. Move the pickup to frame 16,000 (15,000) by scanning or searching. Open (turn off) the tracking servo (remote 3 key). Observe the scope display.

Insert a *special grating adjustment tool* into the grating adjustment hole and turn the grating so that the amplitude of the tracking-error signal varies from maximum to minimum.

Note that only a *special grating adjustment tool* (Philips 0017 168 30019) avail-

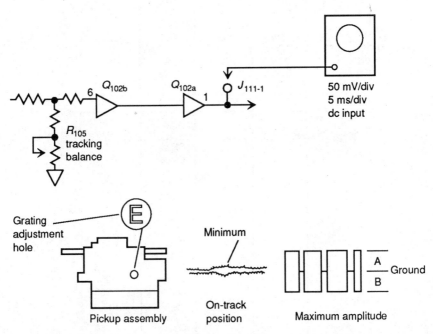

FIGURE 9.2 Rough-grating and tracking-balance adjustments.

able from the player manufacturer can be used. Any other tool may (and probably will) damage the adjustment point. This will usually require replacement of the pickup assembly as a special package (at nerve shattering expense to the customer).

Using the adjustment tool, find the position where the waveform amplitude reaches a minimum with a smooth waveform envelope, as shown in Fig. 9.2. This condition indicates that the three-way split laser beam is directed onto a single track, which is called the *on-track* position.

Slowly turn the grating counterclockwise from the on-track position until the gradually increasing tracking-error waveform reaches maximum amplitude (Fig. 9.2). Close (turn on) the tracking servo (remote 3 key) and check that a normal picture is displayed on the monitor.

Tracking-Balance Adjustment. Leave all connections in their present condition but align the scope trace with the center of the scope screen. Adjust R_{105} so that the positive and negative halves (A and B) of the tracking-error waveform are equal, as shown in Fig. 9.2.

9.3.5 RF Gain Adjustment

Figure 9.3 shows the RF gain adjustment diagram. Figure 6.24 shows additional circuit details. This adjustment sets the RF signal amplitude (from the main pickup photodiodes) to the optimum value.

Play a CDV (LD) test disc. Move the pickup to frame 16,000 (15,000) by scanning or searching using the player keys.

Observe the scope display and adjust R_{282} for an RF signal amplitude of 300 mV ±20 mV, as shown on Fig. 9.3.

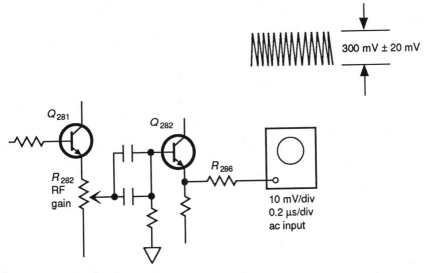

FIGURE 9.3 RF gain adjustment.

9.3.6 Spindle-Motor Centering Check

Figure 9.4 shows the test diagram for the spindle-motor centering check. Figure 6.18 shows additional circuit details. This test checks that a single imaginary line (running parallel to the pickup slide) passes through the center of the pickup laser beam and the center of the spindle.

Play a CD test disc (Philips 5A). Move the pickup to the inner tracks of the disc by scanning or searching. Then open (turn off) the tracking servo (remote 3 key).

Observe the scope display (scope set to X-Y, tracking error to Y and tracking sum to X) and measure the Y axis pattern amplitude, as shown in Fig. 9.4.

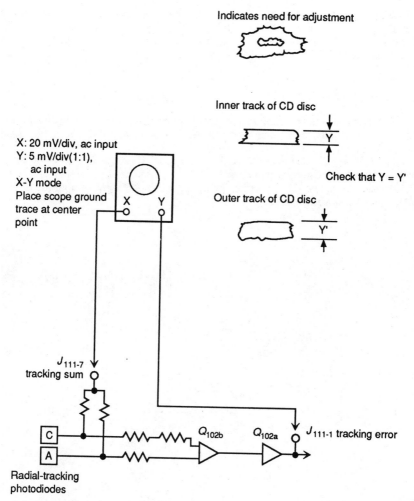

FIGURE 9.4 Spindle-motor centering check.

Turn on the tracking servo and move the pickup to the outer tracks of the disc. Then open (turn off) the tracking servo and measure the Y axis amplitude.

If the amplitude of the Y axis is not substantially the same when the pickup is at the inner and outer tracks of the disc, the spindle-motor centering must be adjusted, as described in Sec. 9.3.7.

Note that this check is required whenever the pickup assembly is replaced.

9.3.7 Spindle-Motor Centering Adjustment

Figure 9.5 shows the spindle-motor centering adjustment diagram. Figure 6.18 shows additional circuit details. This adjustment ensures that the single imaginary line (running parallel to the pickup slide) passes through the center of the pickup laser beam and the center of the spindle.

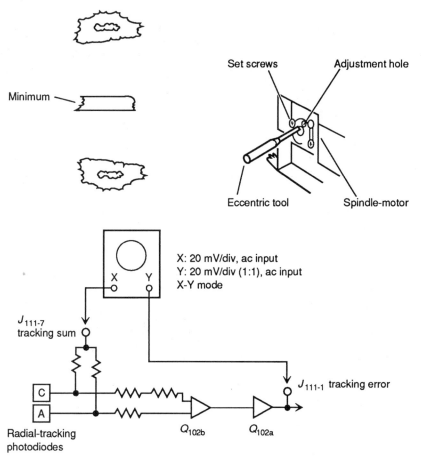

FIGURE 9.5 Spindle-motor centering adjustment.

Note that this adjustment is necessary only when indicated by the spindle-motor centering check, Sec. 9.3.6.

1. Loosen the three spindle-motor set screws by turning each about half a turn.
2. Connect X and Y inputs of the scope as shown in Fig. 9.5.
3. Access the test mode and play a CD test disc (Philips 5A). Move the pickup to the outer tracks of the disc by scanning or searching.
4. Open (turn off) the tracking servo (remote 3 key) and observe the scope display.
5. Using the procedures in Sec. 9.3.4, fine-adjust the grating until the amplitude along the Y axis is minimum.
6. Close (turn on) the tracking servo, and move the pickup to the inner tracks of the disc.
7. Open (turn off) the tracking servo again, and observe the scope display. Record the amplitude of the Y axis.
8. Insert an eccentric tool into the adjustment hole, as shown in Fig. 9.5, and slowly turn in the direction which *reduces* the Y axis amplitude. (The eccentric tool can be fabricated, as shown in Fig. 9.6.) After reaching minimum Y axis amplitude, continue turning the eccentric tool in the *same direction* until the Y axis amplitude is the same as recorded in step 7.
9. Close (turn on) the tracking servo and move the pickup back to the outer tracks of the disc.
10. Repeat steps 4, 5, and 6.
11. Open (turn off) the tracking servo again and observe the scope display. Check that the amplitude of the Y axis has reached a minimum. If the Y axis appears to be greater than minimum, repeat steps 8 through 11. Tighten the spindle-motor set screws.

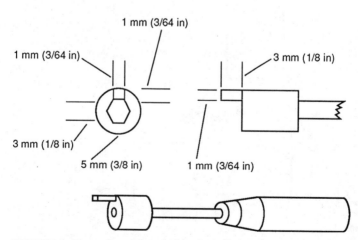

FIGURE 9.6 Eccentric tool for spindle-motor adjustments.

9.3.8 Pickup Tracking-Direction Inclination Adjustment

Figure 9.7 shows the pickup tracking-direction inclination adjustment diagram. Figures 6.17 and 6.23 show additional circuit details. Adjustment of the slider-shaft inclination ensures that the pickup assembly moves parallel to the disc surface. Adjustment of the pickup tracking-direction angle ensures that the laser beam is perpendicular to the disc.

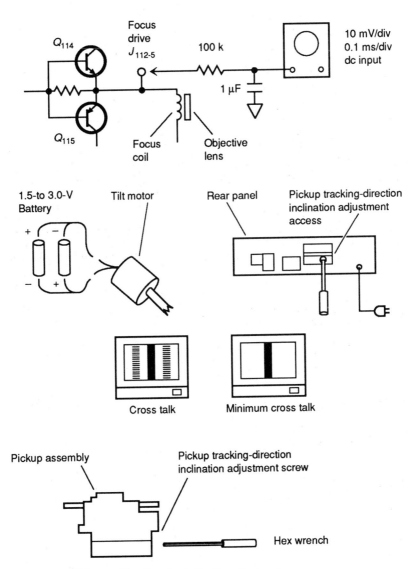

FIGURE 9.7 Pickup tracking-direction inclination adjustment.

1. Disconnect the tilt-motor connector J_{251} (Fig. 6.23). Do not reconnect J_{251} until the procedures of Sec. 9.3.8 through 9.3.11 are complete.
2. Play a CDV test disc, and search to frame 4760 (4760). This is the tilt fulcrum point.
3. Connect the scope to $J_{112\text{-}5}$ through a low-pass filter (100 k, 1 μF), as shown in Fig. 9.7, and observe the focus-drive voltage.
4. Adjust the scope Y axis position knob and move the focus-drive waveform to the center of the scope screen. Measure the focus-drive voltage again.
5. Continue to measure the focus-drive voltage while scanning to frame 46,135 (42,314). If the voltage differs from that measured in step 4, connect a battery (1.5 to 3 V) to the tilt motor (Fig. 9.7), and turn the motor until the focus-drive voltage is within ±50 mV of that in step 4.
6. Remove the small lid on the rear panel, and insert a hex wrench through the opening. Adjust the pickup tracking-direction inclination adjustment screw to minimize the cross talk on the left and right sides of the monitor screen.
7. Search to frame 115 (104) and check that cross talk on the left and right sides of the monitor screen is minimized (and is substantially equal on both sides). Repeat steps 6 and 7 as necessary.

9.3.9 CDV (LD) Focus-Error Balance Adjustment.

Figure 9.8 shows the focus-error balance adjustment diagram. Figure 6.13 shows additional circuit details. This adjustment ensures that the focus servo maintains the objective lens at the optimum distance from the disc during CDV (LD) playback.

Play a CDV test disc, and search to frame 115 (104). Adjust the CDV (LD) focus balance R_{162} to minimize cross talk on both sides of the monitor screen. If this adjustment fails to reduce cross talk down to an allowable level, go to the pickup tangential direction angle adjustment in Sec. 9.3.10. Leave R_{162} at whatever setting produces minimum cross talk.

9.3.10 Pickup Tangential-Direction Angle Adjustment

Figure 9.9 shows the pickup tangential-direction angle adjustment diagram. Figure 6.18 shows additional circuit details. This adjustment is to reduce cross talk. However, the adjustment is usually necessary only if cross talk remains conspicuous after completing the pickup tracking-direction adjustment (Sec. 9.3.8) and the CDV (LD) focus-error balance adjustment (Sec. 9.3.9).

1. Play a CDV (LD) test disc, search to frame 28,600 (27,000), and open the tracking servo.
2. Connect the scope to $J_{111\text{-}1}$ (Fig. 6.18) and measure the tracking-error signal.
3. Insert a hex wrench through the gap between the chassis assembly and mechanical assembly (Fig. 9.9) to the pickup tangential-direction inclination adjustment screw.
4. Adjust the screw until the tracking-error waveform is maximum.
5. Remove the hex wrench. Search to frame 115 (104) and check that cross talk

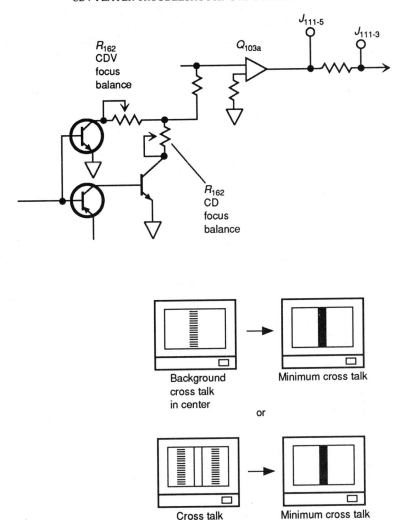

FIGURE 9.8 CDV (LD) focus-error balance adjustment.

on the left and right sides of the monitor is minimized (and is substantially equal on both sides). Repeat steps 4 and 5 as necessary.

9.3.11 Tilt-Sensor Inclination Adjustment

Figure 9.10 shows the tilt-sensor inclination adjustment diagram. Figure 6.23 shows additional circuit details. This adjustment sets the tilt-servo offset voltage to 0 V (by adjustment of tilt-sensor inclination) when the pickup is replaced.

1. Check the color of the dot marked on the flexible cable next to the tilt sensor. There are two types of dots. Adjust tilt gain R_{218} accordingly. For a red dot,

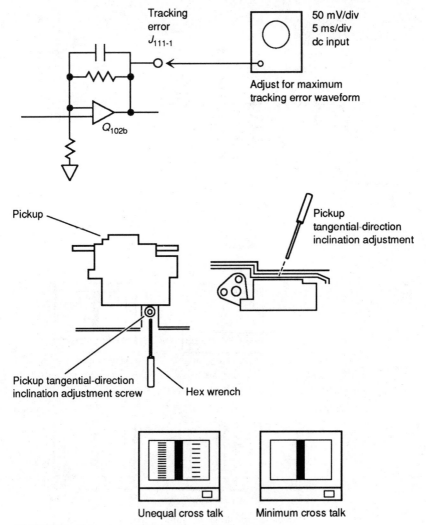

FIGURE 9.9 Pickup tangential-direction angle adjustment.

turn R_{218} fully clockwise. For a blue dot, turn R_{218} fully counterclockwise. If there is no dot or mark, set R_{218} to the center position.

2. Play a CDV (LD) test disc, and search to frame 4760 (4760).
3. Connect the scope to $J_{112\text{-}1}$ and measure the tilt-error dc voltage.
4. Insert a Philips screwdriver with a long shaft through the rear panel (Fig. 9.10) and adjust the tilt-sensor inclination adjustment screw until the tilt-error dc voltage is 0 V.
5. Connect the tilt-motor connector J_{251} that was disconnected during the pickup tracking-direction inclination adjustment (Sec. 9.3.8).

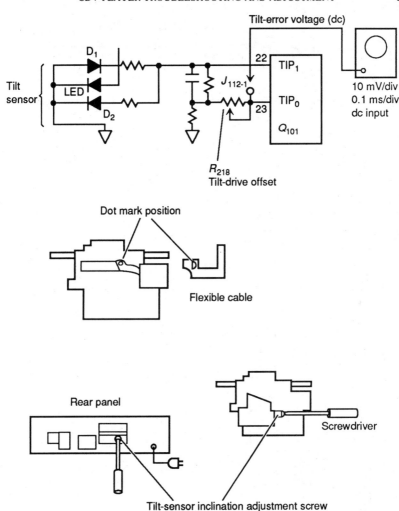

FIGURE 9.10 Tilt-sensor inclination adjustment.

6. Search to frame 115 (104) and check that cross talk on the left and right sides of the monitor has been minimized and is substantially equal on both sides.

9.3.12 Fine-Grating and Tracking-Balance Adjustments

Figure 9.11 shows the fine-grating and tracking-balance adjustment diagram. Figure 6.18 shows additional circuit details. This adjustment (or check) sets the laser beam so that the two tracking beams (on either side of the main beam, Fig. 2.8) are at the optimum position between the disc tracks, and it sets the tracking-servo offset voltage to 0 V.

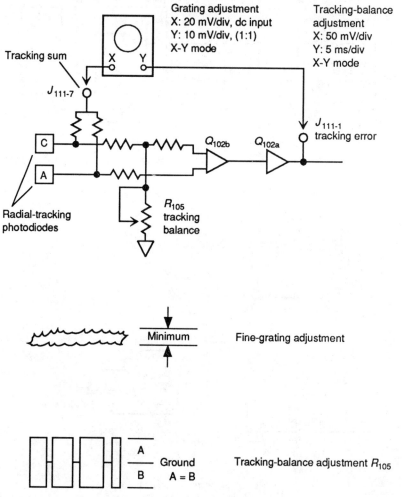

FIGURE 9.11 Fine-grating and tracking-balance adjustments.

1. Access the test mode. Play a CDV (LD) test disc, search to frame 16,000 (15,000), and open the tracking servo.
2. Connect the scope, as shown in Fig. 9.11, and observe the tracking-error and tracking-sum signals.
3. Insert the special grating adjustment tool (Sec. 9.3.4) and fine-adjust the grating until the amplitude of the Y axis reaches minimum. If the grating is turned too far, and the optimum position can no longer be found, repeat the rough-grating adjustment (Sec. 9.3.4).
4. Leave all connections in their present condition, but align the scope trace with the center of the scope screen. Check that the positive and negative halfs (A

CDV PLAYER TROUBLESHOOTING AND ADJUSTMENT   9.17

and B) of the tracking-error waveform are equal, as shown in Fig. 9.11. If they are not, repeat the tracking-balance adjustment (Sec. 9.3.4).

5. Close the tracking servo, and check that a normal picture is shown on the monitor.

9.3.13 Tracking-Servo Loop-Gain Adjustment

Figure 9.12 shows the tracking-servo loop-gain adjustment diagram. Figure 6.18 shows additional circuit details. This adjustment sets the tracking-servo loop-gain to the optimum value.

1. Play a CDV (LD) test disc and search to frame 16,000 (15,000).
2. Connect the resistor, audio generator, and scope to J_{111}, as shown in Fig. 9.12.
3. Set the audio generator to 16 V_{p-p} at 2.73 kHz.
4. Observe the scope pattern with the scope in the X-Y mode.

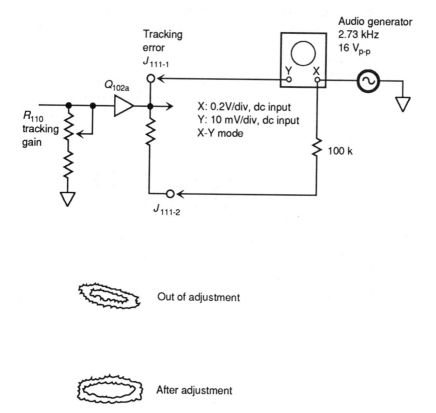

FIGURE 9.12 Tracking-servo loop-gain adjustment.

9.18 CHAPTER NINE

5. Adjust R_{110} until the scope pattern is symmetrical, as shown in Fig. 9.12.

Note that if the audio generator output is not capable of delivering 16 V, it may be necessary to reduce the value of the series resistor between $J_{111\text{-}2}$ and the X input of the scope. Also, if the disc surface is scratched, the waveforms may not be read because of noise.

9.3.14 Focus-Servo Loop-Gain Adjustment

Figure 9.13 shows the focus-servo loop-gain adjustment diagram. Figures 6.13 and 6.17 show additional circuit details. This adjustment sets the focus-servo loop-gain to the optimum value.

1. Connect the gate of field-effect transistor (FET) Q_{116} to ground (Fig. 6.17). This disables the high-frequency focus limiter circuit (Sec. 6.8.6).
2. Connect the resistor, audio generator, and scope to J_{111}, as shown in Fig. 9.13.
3. Set the audio generator to 16 $V_{p\text{-}p}$ at 1.9 kHz.
4. Observe the scope pattern with the scope in the X-Y mode.
5. Adjust R_{168} (Fig. 6.13) until the scope pattern is symmetrical, as shown in Fig. 9.13.
6. Disconnect the gate of Q_{116} from ground.

Note that if the audio generator is not capable of delivering 16 V, it may be necessary to reduce the value of the series resistor between $J_{111\text{-}3}$ and the scope

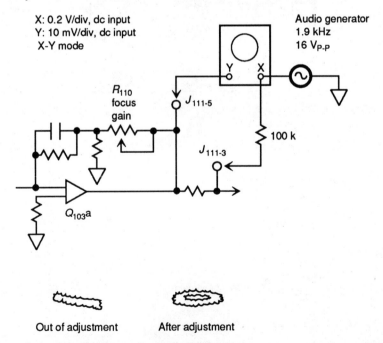

FIGURE 9.13 Focus-servo loop-gain adjustment.

X input. Also, if the disc surface is scratched, the waveforms may not be read because of noise.

9.3.15 CD Focus-Error Balance Adjustment

Figure 9.14 shows the focus-error balance adjustment diagram. Figures 6.13 and 6.25 show additional circuit details. This adjustment ensures that the focus servo maintains the objective lens at the optimum distance from the disc during CD playback.

1. Play a CD test disc (Philips 5A).
2. Connect the scope to $Q_{505\text{-}1}$ (Fig. 6.25), as shown in Fig. 9.14. Observe the EFM signal (so-called eye pattern).
3. Adjust CD focus balance R_{163} until the EFM signal is at maximum.

9.3.16 Focus-Sum Level Adjustment

Figure 9.15 shows the focus-sum level adjustment diagram. Figure 6.13 shows additional circuit details. This adjustment sets the focus-sum level to the optimum value.

1. Play a CDV (LD) test disc and search to frame 4760 (4760).

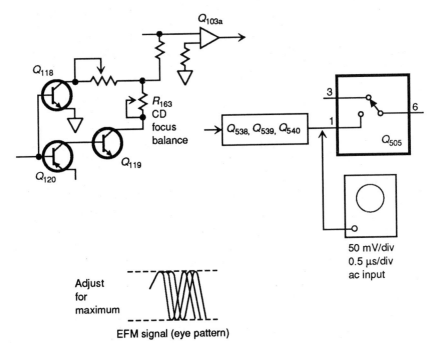

FIGURE 9.14 CD focus-error balance adjustment.

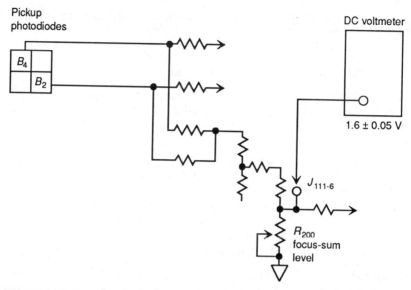

FIGURE 9.15 Focus-sum level adjustment.

2. Measure the focus-sum voltage at J_{111-6}.
3. Adjust R_{200} until the voltage at J_{111-6} is at 1.6 ±0.05V.

Note that it is possible for the player to operate with a focus-sum voltage of less than 1.595 V. However, if the focus-sum voltage is substantially below 1.595 V, this indicates a possible failure of the focus-error circuits (such as Q_{103} and the photodiodes B_1 through B_4, in Fig. 6.13). If the focus-sum voltage is slightly low, set R_{200} for maximum voltage at J_{111-6}, but do not exceed 1.605 V.

9.3.17 Reference Shift Adjustment

Figure 9.16 shows the reference shift adjustment diagram. Figure 6.31 shows additional circuit details. This adjustment sets the amount of reference shift required for detecting a spindle-servo frequency error.

1. Connect the scope to Q_{702-18}, as shown in Fig. 9.16.
2. Adjust the scope trigger control to stabilize the pulse waveform.
3. Adjust reference shift R_{766} until the start of the high portion of the waveform to the start of the low portion is 28 μs ±1 μs.

9.3.18 PLL Offset Adjustment

Figure 9.17 shows the PLL offset adjustment diagram. Figure 6.29 shows additional circuit details. This adjustment sets the offset of 3.58 MHz, which is used in video phase-error detection.

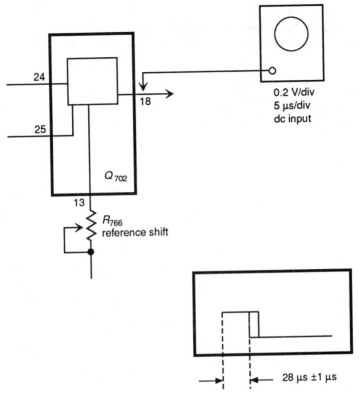

FIGURE 9.16 Reference shift adjustment.

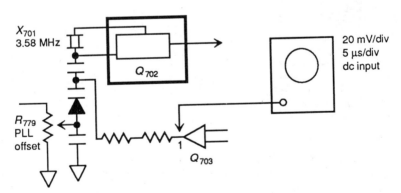

FIGURE 9.17 PLL offset adjustment.

1. Play a CDV (LD) test disc.
2. Connect the scope to Q_{703} as shown in Fig. 9.17.
3. Adjust PLL offset R_{779} until the offset voltage is 0 V ±2 mV.

9.3.19 Half-H Rejection Adjustment

Figure 9.18 shows the half-H rejection adjustment diagram. Figure 6.31 shows additional circuit details. This adjustment sets the MMV pulse width for half-H rejection.

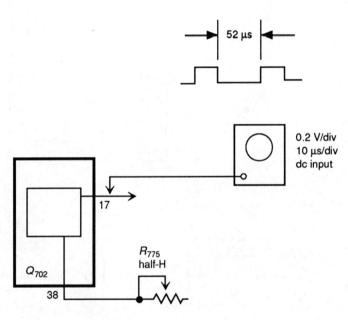

FIGURE 9.18 Half-H rejection adjustment.

1. Play a CDV (LD) test disc.
2. Connect the scope to $Q_{702\text{-}17}$, as shown in Fig. 9.18.
3. Adjust half-H R_{775} until the width of the low interval of the pulse waveform is 52 µs ±1 µs, as shown.

9.3.20 Trapezoid Inclination (TBC Error) Adjustment

Figure 9.19 shows the trapezoid inclination adjustment diagram. Figure 6.31 shows additional circuit details. This adjustment sets the slope of the trapezoidal

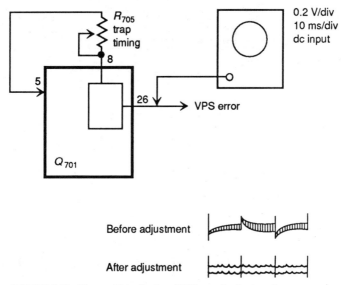

FIGURE 9.19 Trapezoid inclination (TBC error) adjustment.

waveform used in TBC error-detection timing (the video phase shift, or VPS, waveform).

1. Play a CDV (LD) test disc. Put any frame into the still mode.
2. Switch the digital-memory function off.
3. Connect the scope to $Q_{701\text{-}26}$, as shown in Fig. 9.19.
4. Adjust trap timing R_{705} until the VPS error waveform is flat, as shown in Fig. 9.19.

9.3.21 Burst-Gate Position Adjustment

Figure 9.20 shows the burst-gate position adjustment diagram. Figure 6.29 shows additional circuit details. This adjustment sets the position of the burst gate.

1. Play a CDV (LD) test disc.
2. Connect the scope to Q_{527} and $Q_{701\text{-}22}$, as shown in Fig. 9.20.
3. Adjust burst gate R_{709} until the trailing edge of the Memory Video (MMV) output is delayed by about 1 μs ±0.1 μs with respect to the leading edge of the video signal, as shown in Fig. 9.20.

9.3.22 Reference Clock (14.31818 MHz) Frequency Adjustment

Figure 9.21 shows the reference clock frequency adjustment diagram. Figures 6.31 and 6.39 show additional circuit details. This adjustment sets the reference clock frequency.

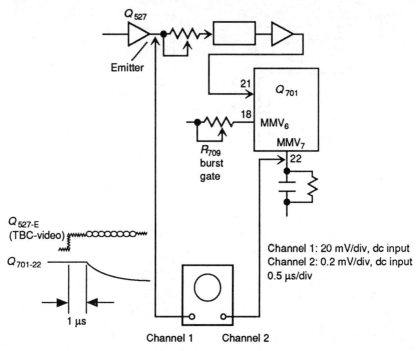

FIGURE 9.20 Burst-gate position adjustment.

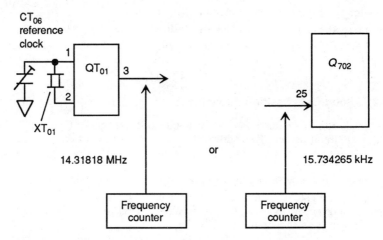

FIGURE 9.21 Reference clock (14.31818 MHz) frequency adjustment.

CDV PLAYER TROUBLESHOOTING AND ADJUSTMENT 9.25

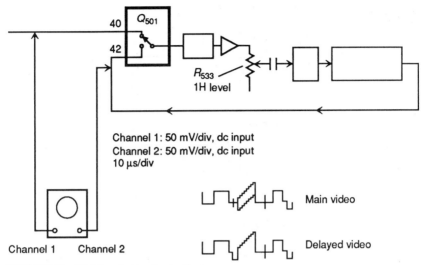

FIGURE 9.22 1H delayed-video level adjustment.

1. Connect a frequency counter to $QT_{01\text{-}3}$, as shown in Fig. 9.21.
2. Apply power to the player, but do not play a disc.
3. Adjust reference clock CT_{06} until the counter reads 14.31818 MHz ± 10 Hz.
4. If this frequency is difficult to adjust (say because the frequency counter does not provide sufficient resolution), play a CDV (LD) test disc, and adjust CT_{06} until the counter reads 15.734265 kHz when connected to $Q_{702\text{-}25}$ (Fig. 6.31).

9.3.23 1H Delayed-Video Level Adjustment

Figure 9.22 shows the 1H delayed-video level adjustment diagram. Figure 6.26 shows additional circuit details. This adjustment sets the amplitude of the 1H delayed-video signal (for dropout compensation) to the same level as the main video signal (so that any dropout is invisible).

1. Play a CDV (LD) test disc and search to frame 19,801 (19,801).
2. Connect the scope to Q_{501}, pins 40 and 42, as shown in Fig. 9.22.
3. Adjust 1H level R_{533} until the amplitude from the sync tip to the white level in the 1H delayed-video signal is the same as the amplitude of the main video signal.
4. When properly adjusted, any dropout will be gray and will be almost invisible on the monitor. If R_{533} is not adjusted properly, the dropout will be visible (black if the delayed level is low or white if the level is high).

9.3.24 VCO Center-Frequency Adjustment

Figure 9.23 shows the VCO center-frequency adjustment diagram. Figure 6.28

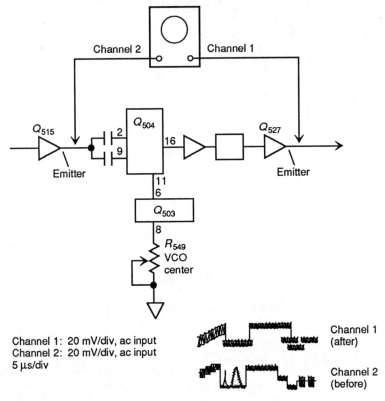

FIGURE 9.23 VCO center-frequency adjustment.

shows additional circuit details. This adjustment optomizes the CCD delay time for time-base error compensation.

1. Play a CDV (LD) test disc and search to frame 19,801 (19,801).
2. Connect the scope to Q_{515} and Q_{527}, as shown in Fig. 9.23.
3. The video signal following the time-base error compensation in channel 1 (emitter of Q_{527}) contains jitter. Adjust VCO center R_{549} to delay the center of the jitter by 70.7 μs (1H + 7.2 ± 0.1 μs) from the trailing edge of the horizontal sync (H-sync) in the video signal prior to time-base error compensation at channel 2, as shown in Fig. 9.23.

9.3.25 Output-Video Level Adjustment

Figure 9.24 shows the output-video level adjustment diagram. Figure 6.28 shows additional circuit details. This adjustment sets the level of the player video output (from pedestal to 100 percent white) to 0.71 V_{p-p}.

1. Connect the player video output terminal to a monitor, and terminate the monitor internally with 75 Ω (if not already so terminated). If you are using a TV

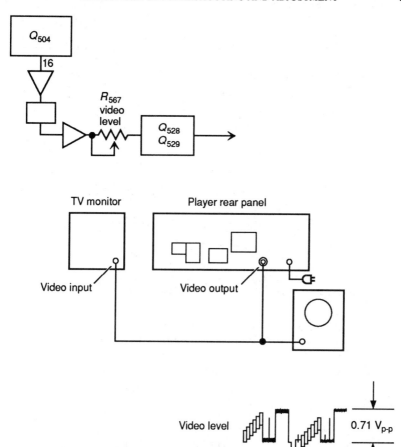

FIGURE 9.24 Output-video level adjustment.

set with no video input terminal, terminate the player video output terminal with 75 Ω.
2. Play a CDV (LD) test disc and search to frame 19,801 (19,801).
3. Connect the scope to the player video output terminal (in parallel with the monitor) and observe the playback video waveform on the scope.
4. Adjust video level R_{567} until the amplitude from the pedestal level to the white level of the playback video waveform is 0.71 V ±5 percent.

9.3.26 Color Phase-Correction Level Adjustment

Figure 9.25 shows the color phase-correction level adjustment diagram. Figure 6.30 shows additional circuit details. This adjustment optomizes the amount of color phase correction.

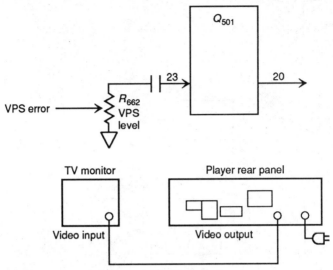

FIGURE 9.25 Color phase-correction level adjustment.

1. Play a CDV (LD) test disc and search to frame 7201 (26,101).
2. Observe the monitor for color streaking, particularly any magenta color irregularities.
3. Adjust the VPS level R_{662} to minimize magenta streaking.

9.3.27 Detector Level Adjustment

Figure 9.26 shows the detector level adjustment diagram. Figure 6.25 shows additional circuit details. This adjustment optomizes the input voltage (spindle error) applied to the spindle-motor speed detector circuit.

1. Play a CDV (LD) test disc and search to frame 4801 (5401).
2. Connect the DVM to Q_{881}, as shown in Fig. 9.26.
3. Adjust detector level R_{883} to obtain a difference voltage of 330 mV ±5 mV.

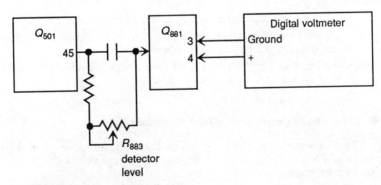

FIGURE 9.26 Detector level adjustment.

9.3.28 Digital-Audio CD-Oscillator Adjustment

Figure 9.27 shows the digital-audio CD-oscillator adjustment diagram. Figure 6.41 shows additional circuit details. This adjustment sets the digital-audio CD clock to the correct frequency.

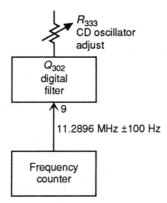

FIGURE 9.27 Digital-audio CD-oscillator adjustment.

1. Connect a frequency counter to Q_{302-9}, as shown in Fig. 9.27.
2. Play a CD test disc (Philips 5A).
3. Adjust R_{333} to set the clock to 11.2896 MHz ± 100 Hz.

9.3.29 Digital-Memory VCXO-Clock Adjustment

Figure 9.28 shows the digital-memory VCXO-clock adjustment diagram. Figure 6.39 shows additional circuit details. This adjustment sets the digital-memory clock to the correct frequency.

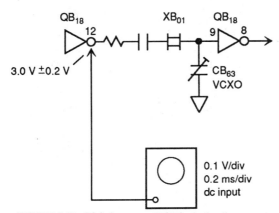

FIGURE 9.28 Digital-memory VCXO-clock adjustment.

9.30 CHAPTER NINE

1. Gain access to the digital-memory unit. Typically, this involves removing the unit from the shield case with the digital-memory assembly inside by taking off the five connecting cables and four screws that hold the shield case to the unit. Then remove the shields and reconnect the five cables.
2. Connect the scope to QB_{18}, as shown in Fig. 9.28.
3. Adjust VCXO CB_{63} to set the voltage at pin 12 of QB_{18} to 3.0 V ±0.2 V.

9.3.30 Sync-Level Adjustment

Figure 9.29 shows the sync-level adjustment diagram. Figures 6.32 and 6.39 show additional circuit details. This adjustment equalizes the amplitude of the H-sync signals of the input video signal applied to digital memory (the "through video" signal) and the H-sync signals of the output video signal from digital memory (the "memory video" signal).

1. Play a CDV (LD) test disc and search to frame 4801 (5401).
2. Connect the scope to J_{502}, as shown in Fig. 9.29. This connection makes it possible to monitor CCD video and memory video simultaneously.

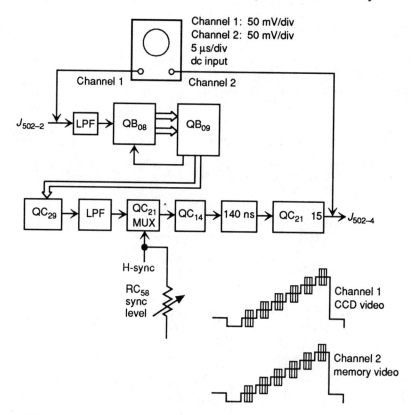

FIGURE 9.29 Sync-level adjustment.

3. Press the player Digital Picture key.
4. Adjust sync-level RC_{58} until the amplitudes of the H-sync signals in digital-memory input and output video signals are the same.

9.3.31 Memory-Video Level Adjustment

Figure 9.30 shows the memory-video level adjustment diagram. Figures 6.32 and 6.39 show additional circuit details. This adjustment equalizes the amplitude of the luma signals of the input video signal applied to digital memory (the through video signal) and the luma signals of the output video signal from digital memory (the memory video signal).

1. Play a CDV (LD) test disc and search to frame 3900 (1000).
2. Connect the scope to J_{502}, as shown in Fig. 9.30. This connection makes it possible to monitor TBC video and memory video simultaneously.
3. Press the player Digital Picture key.

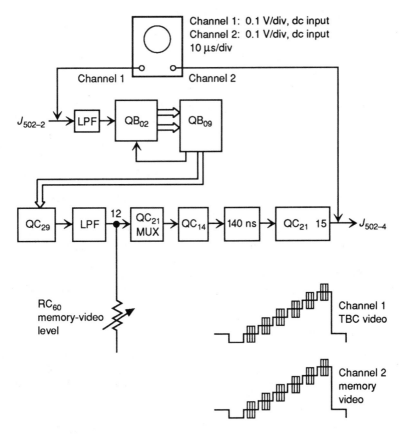

FIGURE 9.30 Memory-video level adjustment.

9.32 CHAPTER NINE

4. Adjust memory-video level RC_{60} until the amplitudes of the luma signals in digital-memory input and output video signals are the same.

9.3.32 Memory-Video (MMV) Adjustment

Figure 9.31 shows the MMV adjustment diagram. Figure 6.39 shows additional circuit details. This adjustment synchronizes the time base of the through video and memory video signals.

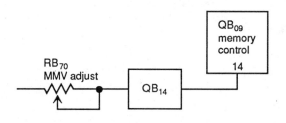

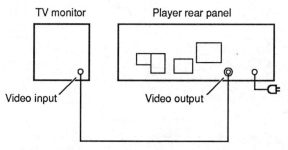

FIGURE 9.31 Memory-video adjustment.

1. Play a CDV (LD) test disc and search to frame 3900 (1000).
2. Press the player Digital Picture key on and off while observing the TV or monitor.
3. Adjust MMV RB_{70} to minimize lateral displacement (horizontal shift) between direct pictures (through video) and pictures passed by the digital memory.

9.3.33 140-ns Shift-Level Adjustment

Figure 9.32 shows the 140-ns shift level adjustment diagram. Figure 6.39 shows additional circuit details. This adjustment sets the amount of phase shift in the memory video 140-ns phase-shift circuit (to provide the correct burst phase to the composite video signal).

1. Play a CDV (LD) test disc and search to frame 7201 (6301).
2. Connect the scope to QB_{17}, as shown in Fig. 9.32.

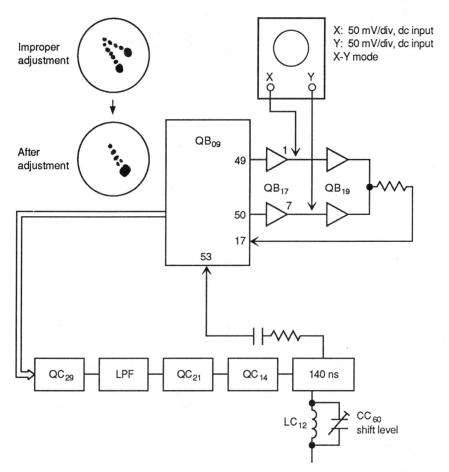

FIGURE 9.32 140-ns shift-level adjustment.

3. Operate the scope in the X-Y mode and observe the scope pattern.
4. Adjust shift level CC_{60} until the two bright points in the scope pattern are as close to each other as possible (overlap is best), as shown in Fig. 9.32.

9.3.34 140-ns Level Adjustment

Figure 9.33 shows the 140-ns level adjustment diagram. Figures 6.32 and 6.39 show additional circuit details. This adjustment equalizes the amplitude of the chroma signal (in the composite video) before and after the 140-ns phase shift.

1. Play a CDV (LD) test disc and search to frame 7201 (6301).
2. Connect the scope to $QC_{21\text{-}15}$, as shown in Fig. 9.33.
3. Adjust 140-ns level RC_{59} to minimize vertical deviation of the chroma signal (as shown on the scope).

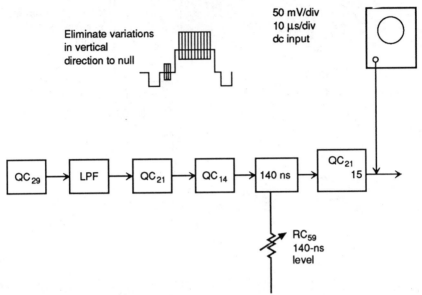

FIGURE 9.33 140-ns level adjustment.

4. Observe the magenta image on the monitor screen to check for minimum flicker of the magenta. If excessive magenta flicker persists, repeat the 140-ns shift-level adjustment (Sec. 9.3.33). If a vectorscope is available, adjust RC_{59} to align the magenta gain in the vectorscope screen to the prescribed position (typically 61° from B-Y, chroma 80 IEEE units, and luma 36 IEEE units).

9.3.35 Y/C Separation Adjustments

Figure 9.34 shows the Y/C separation adjustment diagram. Figure 6.38 shows additional circuit details. These adjustments provide for the correct separation of Y and C signals from the composite video signal.

1. Remove the Y/C separation circuit board.
2. Play a CDV (LD) test disc and search to frame 9500.
3. Connect the scope to pin 15 of PS_{26} and observe the CCD Out waveform (after passing through the LPF). Adjust CCD Output R_{12} to minimize distortion and maximize the level simultaneously. Use a 0.1-μs/div sweep on the scope.
4. Search to frame 34,000 and observe the chroma-output (C-Out) waveform. Adjust C-Phase R_{24} and C-Level R_{20} alternately to minimize the residual Y component at 3.58 MHz. Use the luma-output (Y-Out) on channel 2 of the scope as a reference for location of the 3.58-MHz burst.
5. Search to frame 9500 and observe the luma-output (Y-Out) waveform. Adjust Y-Phase R_{38} and Y-Level R_{55} alternately to minimize the residual C component.
6. Search to frame 6500 and check that the 100 percent level is 714 ±55 mV. If

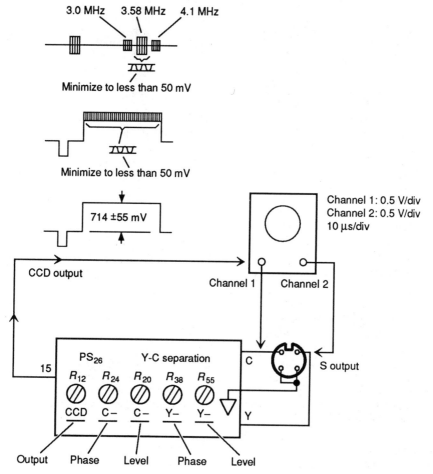

FIGURE 9.34 Y/C separation adjustments.

the level is not correct, repeat the video output-level adjustment (Sec. 9.3.25) as necessary.

9.4 CDV TROUBLE SYMPTOMS RELATED TO ADJUSTMENT

The following notes describe trouble symptoms that can be caused by improper adjustment. Remember that these same symptoms can also be caused by circuit failures. (We describe circuit failure as the basis of trouble symptoms in the remaining sections of this chapter.)

Improper tracking (track jumping or skipping): Although they are not the only possible causes, these problems are often caused by improper grating and tracking-balance adjustments (Secs. 9.3.4 and 9.3.12) or the focus-sum level adjustment (Sec. 9.3.16).

Frequent dropouts: If there are frequent dropouts (with more than one disc), start by checking the RF-gain adjustment (Sec. 9.3.5). If the RF signal is weak, even a properly functioning DOC circuit cannot function properly.

Excessive dropouts: Once you are certain that RF gain is good (Sec. 9.3.5), check the 1H delayed-video level adjustment (Sec. 9.3.23).

Track jumping during long search times: If tracking is good between nearby tracks but tends to jump when there is a long search time, check the spindle-motor centering (Secs. 9.3.6 and 9.3.7).

Crosstalk on TV or monitor display: If the TV or monitor display shows repeated images or blending of colors (left, right, or both), check the pickup tracking-direction inclination adjustment (Sec. 9.3.8), the CDV focus-error balance adjustment (Sec. 9.3.9), and the tilt-sensor inclination adjustment (Sec. 9.3.11). *If cross talk is excessive*, check the pickup tangential-direction angle adjustment (Sec. 9.3.10).

Degraded playability: If more than one known-good disc (both CDV and CD) show poor quality (both video and audio), check the tracking-servo loop-gain adjustment (Sec. 9.3.13) and the focus-servo loop-gain adjustment (Sec. 9.3.14).

Noise in CD playback sound: If there is noise only in the CD sound, check the focus-error balance adjustment (Sec. 9.3.15).

Spindle-servo locking failure: If the spindle-motor runs but fails to lock in any mode, check the reference shift adjustment (Sec. 9.3.17).

Color irregularities: If the color is poor (or incorrect) on more than one disc, check the PLL offset adjustment (Sec. 9.3.18). If there are color irregularities, together with other symptoms, check as described in the following notes. *If color irregularities are excessive*, check the color phase-correction level adjustment (Sec. 9.3.26).

Picture irregularities and spindle locking failure: If these symptoms occur simultaneously, check the half-H rejection adjustment (Sec. 9.3.19) and the detector level adjustment (Sec. 9.3.27).

Flickering in still mode and displacement in memory pictures: The first place to check for flickering or horizontal displacement in memory pictures is the trapezoid inclination (TBC error) adjustment (Sec. 9.3.20).

Playback does not start at the start of the disc. Missing or irregular color, fine stripes: Start by checking the burst-gate position adjustment (Sec. 9.3.21).

Color irregularities and spindle locking failure: If these symptoms occur simultaneously, check the reference clock (14.31818 MHz) frequency adjustment (Sec. 9.3.22).

Color-lock failure or slow color lock after search: Start by checking the VCO center-frequency adjustment (Sec. 9.3.24).

TV or monitor screen too dark or bright. Replay started at wrong position because of misread data: Start by checking the output video-level adjustment (Sec. 9.3.25).

Digital sound interrupted or irregular: Start by checking the digital-audio CD-oscillator adjustment (Sec. 9.3.28).

Irregular horizontal sync during digital memory mode: Start by checking the digital-memory VCXO clock adjustment (Sec. 9.3.29).

Unstable memory video playback but good nonmemory video: Start by checking the sync-level adjustment (Sec. 9.3.30).

Variations in brightness of the memory video playback but good nonmemory video: Start by checking the memory-video level adjustment (Sec. 9.3.31).

Horizontal displacement of memory video picture with respect to the nonmemory video: Start by checking the MMV adjustment (Sec. 9.3.32).

Flickering on TV or monitor display in all modes: Start by checking both the 140-ns shift-level (Sec. 9.3.33) and 140-ns level (Sec. 9.3.34) adjustments.

S-Video monitor display problems but good nonseparated video: If video from VHF Out and Video Out is good but S-Video is not good (striped noise, images not colored, etc.), check the Y/C separation adjustments (Sec. 9.3.35).

9.5 POWER-SUPPLY TROUBLESHOOTING APPROACH

Figures 9.35 through 9.37 show the circuits involved in power-supply troubleshooting. Since the power supplies provide several voltage sources, a failure in

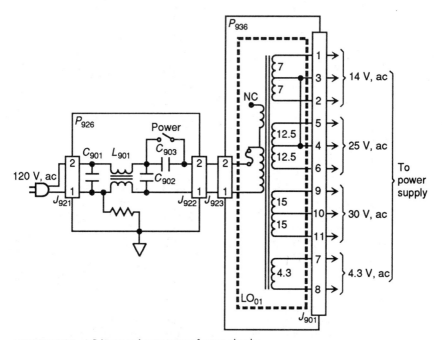

FIGURE 9.35 AC input and power transformer circuits.

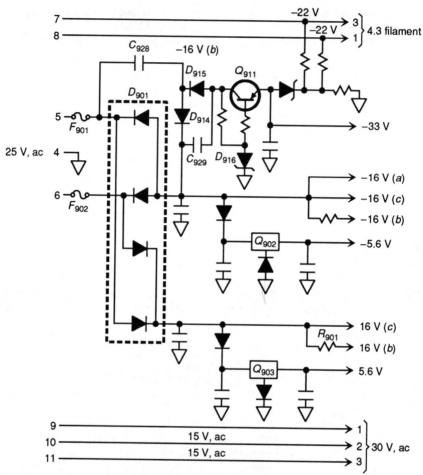

FIGURE 9.36 AC and dc power-supply circuits.

any of them can produce a number of trouble symptoms. Likewise, a problem in the circuits being powered by the supply can make the supply appear defective. However, since none of the supply circuits are that unique, conventional troubleshooting techniques can be used.

As an example, if symptoms indicate a possible defect in a particular power supply, the obvious approach is to measure the dc voltage. If there are many branches, measure the voltage on each branch. If any of the branches are open (say an open R_{901} in Fig. 9.36), the voltage on the other branches may or may not be affected. On the other hand, if any of the branches are shorted, the remaining branches are usually affected (the output voltage is lowered).

If a current probe is available, measure the current in each branch. Most CDV power supplies use PC boards, so you cannot disconnect each output branch (in turn) until a short or other defect is found.

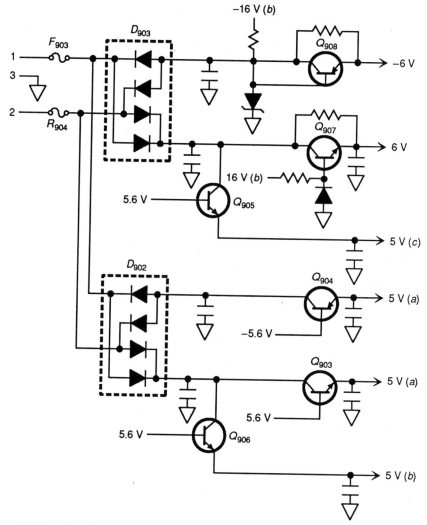

FIGURE 9.37 DC power-supply circuits.

If one or more output voltages appear to be abnormal and it is not possible to measure the corresponding current, remove the power and measure the resistance in each branch. Compare actual resistances against those in the service literature (if available). If you have no idea as to the correct resistance, look for *obvious low resistance* (a complete short or resistance of only a few ohms).

Always use an isolation transformer for the ac power line when checking CDV power supplies (or any electronic equipment for that matter).

If it is necessary to replace a zener diode in any of the regulator circuits, use an *exact replacement*. Some technicians replace zeners with a slightly different voltage value and then try to compensate by adjustment of the regulator. This may or may not work. There are no adjustments in our power-supply circuits (but this is not always true for all CDV players).

Do not overlook the use of scopes in troubleshooting power-supply circuits. Even though you are dealing with dc voltages, there is always some *ripple* present. The ripple frequency and waveform produced by the ripple can help in localizing possible troubles. It is also possible to use a scope as a voltmeter (although if you are only to measure voltage, the meter is simpler).

9.5.1 16-V Power Supply Troubleshooting

If all of the 16-V sources (both positive and negative) are missing, the obvious suspects are fuses F_{901} and F_{902} and D_{901} (Fig. 9.36). Note that if the 16-V sources are missing, the regulated positive and negative 5-V sources will also be missing, since the 5-V sources are derived from the 16-V sources. However, if only the 16-V (b) source is missing, R_{901} is the likely suspect.

9.5.2 5-V Power Supply Troubleshooting

If the 5-V (a) source is missing, the source voltage and bias source for the 5-V (a) voltage regulator should be checked. If both sources are present, check the pass regulator Q_{903}.

9.5.3 33-V Power Supply Troubleshooting (Fluorescent Display)

A failure in the 33-V power supplies can cause problems with the front-panel fluorescent-tube display. The first place to check is at the collector of Q_{911}. If the voltage (approximately -45 V) is present at the collector, suspect Q_{911} or D_{916}. If the voltage at the collector of Q_{911} is absent or abnormal, suspect the voltage-tripler circuit D_{915} and D_{914} or C_{928} and C_{929}.

9.5.4 Power-Supply Source Troubleshooting

Note that there are several regulated sources (series or pass regulation) in our CDV player. Many of the regulated sources are developed by more than one active device and source. For example, the 5-V (a) source is derived from the rectified 14-V source, through pass regulator Q_{903}. However, the bias for Q_{903} is provided by the 5.6-V regulated source through Q_{901}. Therefore, a fault in the 5.6-V source affects the 5-V regulated source.

9.6 SPINDLE-DRIVE TROUBLESHOOTING APPROACH

Figures 6.4 and 6.5 show the circuits involved in spindle-drive troubleshooting. Several symptoms can appear if a fault exists in the spindle-drive circuits.

If the spindle motor does not spin with any disc, check the voltage sources to the drive circuits, particularly the 16-V (c) and 5-V (a) sources. If either voltage is absent, check the power-supply circuits (Sec. 9.5). If both voltages are present, check at pin 9 of the spindle-motor control QD_{01} for a 5-V start signal. If it is absent, suspect the system-control circuits (Sec. 9.7). If the signal is present, suspect QD_{01}.

If the spindle motor spins when a CD is played, the fault may be in the spindle motor or the motor drivers. If there is a fault in either circuit, the audio from a CD will be very poor because of dropouts.

To verify that a fault exists in the spindle motor, check for continuity through the three phases of the motor with an ohmmeter. If each phase of the motor is good (about 4 Ω), suspect the spindle-motor drivers QD_{02} and QD_{03}. The exact resistance of each winding is not critical, but all three windings should show substantially the same resistance.

It is also possible that the spindle drivers are not receiving signals from QD_{01} or a spindle-drive signal, as shown in Fig. 6.5. If any of these signals are absent, check back to the source (Figs. 6.33 through 6.36).

9.7 SYSTEM-CONTROL TROUBLESHOOTING APPROACH

Figures 6.6 through 6.9 show the circuits involved in system-control troubleshooting. A malfunction in any of the CDV systems may occur if there is a failure in the system-control circuits. Since there are several microprocessors involved, the symptoms may point to only one of the microprocessors, but do not count on it. More likely, operation of one system-control microprocessor depends on bus signals from another microprocessor (as is typical for most present-day electronic devices).

Before you start looking for missing inputs to a system-control microprocessor or for outputs from a microprocessor (in response to an input), there are some points that can be checked on any microprocessor system.

9.7.1 Power and Ground Connections

The first step in tracing problems on a suspected system-control IC is to check all power and ground connections. Figures 6.6 and 6.7 show the power and ground connections for the system-control ICs of our CDV player.

Make certain to check *all power and ground* connections to each IC since many ICs have *more than one power and one ground connection*. For example, as shown in Fig. 6.6, pins 1 and 42 of QU_{01} are both grounded. Likewise, as shown in Fig. 6.7, pins 10, 25, and 42 of QU_{03} are grounded.

9.7.2 Reset Signals

With all of the power and ground connections confirmed, check that all of the ICs are properly set. The reset connections for the ICs are shown in Figs. 6.6 and 6.7. Figure 6.8 shows the reset circuit details.

One simple way to check the reset functions is to check for reset pulses at the appropriate pins. For example, pin 6 of QU_{01} (and the entire reset, or $\overline{RES}$, line) should show pulses, as illustrated in Fig. 6.8. If they do not, suspect QU_{16}, QU_{17}, DU_{03}, and CU_{11}.

As in the case of any microprocessor-controlled device, if the reset line is open, shorted to ground, or to power (5 or 16 V), the ICs are not reset (or remain locked in reset) no matter what control signals are applied.

This brings the entire CDV player operation to a halt. So if you find a reset line (or reset pin) that is always high, always low, or apparently connected to nothing (floating), check the line carefully.

9.7.3 Clock Signals

As shown in Fig. 6.6, there are three clock signals in the system-control circuits of our CDV player. The main system-control IC QU_{01} has a 4-MHz clock at pins 2 and 3, keys/display IC QF_{01} has a 6-MHz clock at pins 19 and 20, and the digital-audio IC Q_{304} has a separate 4-MHz clock at pins 15 and 16.

It is possible to measure the presence of a clock signal with a scope or probe. However, a frequency counter provides the most accurate measurement. Obviously, if any of the ICs do not receive the clock signal, the IC cannot function. On the other hand, if the clock is off frequency, all of the ICs may appear to have a clock signal, but the IC function can be impaired. (Note that crystal-controlled clocks do not usually drift off frequency but can go into some overtone frequency, typically a third overtone.)

If the main system-control 4-MHz clock is absent or abnormal, suspect XU_{01}. If the digital-audio 4-MHz clock is absent or abnormal, suspect X_{302}. If the 6-MHz clock is absent or abnormal, suspect XF_{01}.

9.7.4 Bus Circuits

Generally, if there is pulse activity on the buses that interconnect system-control ICs (or any ICs), it is reasonable to assume that the clock, reset, power, and ground are all good. While it is not practical to interpret the codes passing between ICs on the system buses, the presence of pulse activity *on all of the bus lines* indicates that the system-control functions are normal (but do not count on it).

Once you have checked all of the bus lines for pulse activity, the next step is to measure the resistance to ground (with the power off) of each line in the bus (data bus, address bus, control bus, etc.). Pay particular attention to any lines that do not show pulse activity. If any one line differs substantially, suspect a problem on this line. If two lines show the same (higher or lower) resistance, the two lines may be shorted together. In any case, before going further, check the schematic to see if the circuits connected to suspect lines could explain the differences.

9.8 SERVO TROUBLESHOOTING APPROACH

Several symptoms can show up if there is a fault in one of the servo circuits (Fig. 6.12). Such symptoms may include *no start-up* or *poor play* (*poor tracking* or *skipping*), although these symptoms are not exclusively related to the servo.

The most practical approach is to try localizing the problem to only one of the servo systems. The test modes described in Sec. 9.3.3 can be used to help with the isolation process. Also, if you run through the test and adjustment procedures described in Sec. 9.3 (using the recommended test discs), specific problems can be isolated to a particular servo. The following paragraphs describe some techniques for isolating problems in the servo circuits.

9.8.1 Player Will Not Play Any Disc

A likely cause for a player will not play any disc symptom is a problem in the start-up circuits (Figs. 6.9 through 6.17). The start-up sequence can be checked by observing the pickup drive mechanism when ac power is applied.

With the top removed and without a disc, the pickup-drive mechanism should move outward, and the loading tray should move up and down to perform the disc-clamping sequence. After the loading tray moves into the down position, the objective lens should move up and down to perform a focus search.

Since the laser beam cannot focus without a disc, the pickup-drive mechanism moves to check for the presence of a small disc (CD or CDV Single). The focus-search pattern is repeated after reaching the CD focus-search position. If this sequence is not followed, check the focus-servo circuits as follows.

With no disc installed, place the player in a test mode and monitor the signal at J_{112-5} (Figs. 6.10 and 6.17) with a scope. With the test mode initiated, the voltage should rise to 1 V, down to -1 V, and back up to 0 V (producing a voltage ramp) in about 1 s. This pattern should be repeated five times.

If the signal is present at J_{112-5} but the objective lens does not move up and down five times, suspect the laser-diode assembly. If the focus start-up signal is not present, suspect Q_{101}, Q_{104}, Q_{114}, and Q_{115}. Note that the UPDN (start-up) signal at pin 36 of Q_{101} should be about 0.7 V_{p-p}.

If the start-up sequence is functioning, check the laser output next. Monitor the voltage at pin 5 of Q_{108} (Fig. 6.10) and initiate a test mode. While the objective lens moves up and down, pin Q_{108-5} should read about 5 V.

If the 5 V are not present, check the LD On signal from QU_{03-62}. If the 5 V are present, monitor the output at Q_{108-1} and initiate the test mode. Pin 1 of Q_{108} should go to about 2.5 V. If not, suspect Q_{108}.

If Q_{108-1} is 2.5 V with the test mode initiated, check the monitor diode input at pin 3 of Q_{108}. Q_{108-3} should be -5 V when the laser diode is off. When the test mode is activated, Q_{108-3} should go to about -3.6 V. If the voltage does not change when the test mode is initiated, suspect the laser-diode assembly.

9.8.2 Player Operates but Play Is Poor

If the start-up circuits are functioning properly, all of the drives for the servo circuits should be good. However, the adjustment may be off just enough to cause play problems. An effective troubleshooting procedure is to perform the adjustment checks for each of the servo circuits, as described in Sec. 9.3. Section 9.4 relates the procedures to trouble symptoms.

By simply going through the adjustment procedures (in the proper sequence) you can often pinpoint servo-system troubles. For example, if you are adjusting the focus-servo loop-gain, as described in Sec. 9.3.14, and find the signals at $J_{111\text{-}5}$ and $J_{111\text{-}3}$ (Fig. 6.13) absent or abnormal, you have localized trouble to the photodiodes B_1 through B_4 or to the Q_{103} circuits. Note that it may be difficult to monitor the outputs directly from the photodiodes. However, the voltage at all four photodiode test points $J_{113\text{-}4}$ through $J_{113\text{-}1}$ should be substantially the same.

Another aid in troubleshooting the servo circuits is to play different types of discs. For example, if a CD or the audio part of a CDV Single plays properly but a CDV shows poor play, the problem can be isolated to a circuit common to the CDV servo only (such as the CDV focus balance R_{162} or to Q_{118}, Fig. 6.13).

9.9 VIDEO AND TBC TROUBLESHOOTING APPROACH

Figures 6.27 through 6.32, and 6.38 show the circuits involved in video and TDC troubleshooting. There are many different symptoms that can be caused by a failure in the video and TBC circuits. These include *no picture, noisy picture*, and *picture jitter*. Start by playing a test disc and observing the display. This can often pinpoint the problem. Listening to the spindle motor can also give a clue as to the problem. The following techniques will aid in isolating problems in the video and TBC circuits.

9.9.1 No Picture

It is possible that a no-picture symptom can be caused by the spindle motor not being locked. If the spindle sounds like the motor is searching (increasing and decreasing spindle-motor velocity), the problem is probably in the TBC error-detection circuits (Fig. 6.31).

Another indication of TBC-circuit failure is a lack of audio. So if you get no picture and no audio (or very poor audio) but the spindle motor is running, check the TBC error-correction circuits (Figs. 6.27 through 6.32).

One way to check the TBC circuits from input to output is to perform the adjustment checks for each of the circuits, as described in Sec. 9.3. By going through the procedures (in the proper sequence) you can often pinpoint TBC circuit problems.

Section 9.4 relates the adjustment procedures to trouble symptoms. For example, Sec. 9.4 lists our TBC-related trouble symptom of "spindle servo locking

CDV PLAYER TROUBLESHOOTING AND ADJUSTMENT   9.45

failure" as possible improper adjustment of the reference shift R_{766} (described in Sec. 9.3.17).

9.9.2 Poor Video

If the spindle motor appears to be locked, there is good audio, but the video is poor (jitter, noise, etc.) or the video is totally absent, the circuits following the TBC block are suspect. These circuits are shown in Figs. 6.37 and 6.38.

Play a test disc and monitor the circuit signals with a scope. As usual, use the waveforms shown in Figs. 6.37 and 6.38 for reference only. (This applies to any of the simplified schematics in this book.) The service-literature waveforms should provide much greater accuracy (particularly if they are in the form of photos).

If poor video is in the form of flickering on the TV or monitor in all modes, check the 140-ns level adjustments described in Sec. 9.3.34.

If video is poor only when an S-Video (S-VHS) monitor is used, check the Y/C-separation adjustments described in Sec. 9.3.35.

9.10 AUDIO TROUBLESHOOTING APPROACH

Figures 6.24, 6.25, and 6.40 through 6.42 show the circuits involved in audio troubleshooting. This involves straightforward signal tracing. There are no adjustments. Obviously, if the problem is with analog audio but there is good digital audio, check the circuits in Fig. 6.40. On the other hand, check the circuits in Figs. 6.41 and 6.42 if the problem is in digital audio but there is good analog audio.

Remember that both audio outputs must pass through relay L_{402} (Fig. 6.41). If both digital and analog audio are good up to L_{402} but not at the output jacks, suspect L_{402}, Q_{415}, and Q_{304}. Also remember that both digital and analog audio originate from the same source as video (the B_1 through B_4 photodiodes, Fig. 6.24) but are separated in the RF signal-processing circuits, Fig. 6.25.

For example, if video is good, it is reasonable to assume that the photodiodes Q_{281}, Q_{282}, Q_{501}, Q_{511}, and Q_{512} are good. If only analog audio is bad (good video and digital audio), suspect F_{503}, Q_{546}, Q_{547}, and the circuits in Fig. 6.40.

If only digital audio is bad (good video and analog audio), suspect Q_{538}, Q_{539}, Q_{540}, Q_{536}, F_{502}, Q_{537}, Q_{501}, Q_{505}, and the circuits in Figs. 6.41 and 6.42.

9.11 DIGITAL-MEMORY (SPECIAL EFFECTS) TROUBLESHOOTING APPROACH

Figure 6.39 shows the circuits involved in digital-memory troubleshooting. A failure in the digital-memory circuits usually occurs *only* when special effects are selected. Rarely will a problem in digital memory affect normal video.

If the symptom is no special effects, check the video signal input to analog-to-digital (A/D) converter QB_{08}. If the input is absent or abnormal, trace back to the source. If there is a good input to QB_{08}, select the memory mode and play a test disc. Then trace through the circuits in Fig. 6.39.

When a picture is displayed (such as a color-bar test pattern), press the Memory key on the remote control. MEM.P should be displayed on the screen for 1s, indicating storage of the picture being displayed. Now press the Stop key. The player should go into the stop mode, but the memory picture should still be displayed.

If the memory picture is not displayed, check the output from pin 6 of digital-to audio control (DAC) QC_{29}. If the signal is present, trace through the circuits from QC_{29} to pin 15 of QC_{21}. Remember that the 140-ns phase-shift circuits require a 5-V CINV signal from pin 60 of QB_{09} to produce the correct output.

If the signal at pin 6 of QC_{29} is absent or abnormal, check the 8-bit bus input to QC_{29}. Again, it is not practical to interpret the 8-bit code from QB_{09} to QC_{29}. However, all of the bus lines should show pulse activity (and all should show the same resistance to ground) when checked, as described in Sec. 9.7.4).

If one bus line shows no activity, that line is suspect. If all lines show no activity when the memory mode is selected, suspect QB_{08}, QB_{09}, QB_{10}, and the related circuits (edge detect, clock, etc.). Also try performing all of the memory adjustments described in Sec. 9.3.

If the symptom is poor special effects (tearing, loss of sync, distorted video, etc.) but there is good normal video, check all of the circuits in Fig. 6.39. Pay particular attention to all of the inputs from other circuits. For example, if the horizontal sync is missing during memory, check for proper H-sync signals at pin 13 of QC_{21} and for proper adjustment of sync level RC_{58} (Sec. 9.3.30).

9.12 HELIUM-NEON LASER TROUBLESHOOTING AND ADJUSTMENT

Figures 6.43 and 6.44 show the circuits involved in helium-neon laser troubleshooting. *Warning*: Extreme care must be exercised while working with any helium-neon laser supply circuits to prevent the possibility of damaging the regulator circuits or exposing yourself to dangerous voltages.

The helium-neon laser supply should not normally require checking. If the laser is on and 5 V ±0.1 V is measured at TP_5 (Fig. 6.44), the laser supply is operating normally. Likewise, the supply should not normally require adjustment. However, use the following procedures if it becomes necessary to check and/or adjust the laser supply during service.

9.12.1 Checking a Helium-Neon Laser Supply

1. Turn off the power.
2. Disconnect the laser.
3. Ground TP_4 (Fig. 6.44).
4. Close the player lid. Turn the power on.
5. Use a meter capable of handling −2500 V. Connect the meter from the laser terminals to ground. The voltage should be −2000 V ±500 V and typically about −1800 V.
6. Disconnect the meter. Remove the ground lead. Turn the power off and reconnect the laser.

9.12.2 Laser Current Adjustment

1. Close the lid and start the player.
2. Connect a dc voltmeter to TP_5 (Fig. 6.44).
3. Adjust laser-current adjust R_2 for 5 V ±0.1 V.

9.12.3 Helium-Neon Laser Circuit Troubleshooting

1. To determine if the laser is conducting, check for 5 V at TP_5.
2. If 5 V is present at TP_5, the laser is conducting and the laser supply is operating normally. (Most CDV players with helium-neon lasers have some similar test point for a quick check of the laser that does not require measurement of the laser high-voltage supply.)
3. If 5 V is missing at TP_5, check for 25-V_{p-p} drive pulses at TP_6.
4. If the pulses are present, skip to step 8.
5. If the pulses are missing at TP_6, check for a 12-V switched power source.
6. If 12 V is missing, check the lid switch (and any other interlocks used in the player being serviced).
7. If 12 V is present but there are no pulses at TP_6, check Q_2, Q_3, Q_4, and Q_5.
8. If pulses are present at TP_6, check for 8 V at the base of Q_1.
9. If the 8 V is low or missing, check Z_1, D_5, D_6, Q_1, R_1, and R_{25}.
10. If 8 V is present, check the laser high-voltage circuit (Fig. 6.43).
11. If the laser high voltage (about 1800 V) is missing, check all components in Fig. 6.43.
12. If the laser high voltage is present, check all components in Fig. 6.44, including adjustment of R_2.

Index

Adjustments:
  CD, **8**.5 to **8**.11
  CDV, **9**.3 to **9**.35
  dropouts, **8**.9
  microswitch, **7**.13, **7**.25
  turntable, **8**.9
AF (*see* Automatic focus)
Alignment discs, **4**.7
Analog audio, **6**.44
Antishock circuits, **5**.18
Automatic focus (AF), **1**.9, **5**.11, **6**.3, **6**.12, **6**.17
  troubleshooting, **8**.18

Belt, replacement, **7**.8, **7**.23
Bilingual audio, **6**.46
Bit stream, **1**.11
Breakdown, electrostatic, laser, **4**.5

CAV standard play, **1**.19
CD directory, **1**.15
CD-I, **1**.1
CD-ROM, **1**.1, **5**.21 to **5**.28
CDV, **1**.16
CDV single, **1**.21
CED, **1**.2
CIRC, **2**.4
Circuits, antishock, **5**.18
Clock, microprocessor, **6**.11
CLV extended play, **1**.11, **1**.12, **1**.19
Condensation, optics, **4**.2
Control codes, CDV, **1**.18
CX noise reduction, **1**.22, **6**.46

Data processor, **1**.10
Data strobe, **1**.10
Decoding, CD, **2**.8
Deinterleaving, **1**.10, **2**.9
Demultiplexer, **2**.10

Digital:
  audio, **6**.46
  filter, **2**.10
  special effects, **1**.23, **6**.43
  troubleshooting, **9**.45
Directory, CD, **1**.15
Disc care, **4**.9
Disc detector, **7**.28
Disc directory, **1**.15
Distortion meters, **4**.8
DOC, **6**.30
Dropout adjustments, **8**.9

EFM (eight-to-fourteen modulation), **2**.6, **4**.5
Electrostatic breakdown, laser, **4**.5
Encoding, CD, **2**.1
ERCO, **2**.8
Error correction, CD, **2**.4, **2**.8
Eye pattern (HF), **5**.8

Focus, **2**.12, **2**.16, **5**.11, **6**.3, **6**.12, **6**.17 to **6**.21
  adjustments, **8**.9
  automatic, **1**.9, **5**.11, **6**.3, **6**.12, **6**.17
  troubleshooting, **8**.18
Focus, CDV, **6**.3
Frames, CD, **2**.5

Gold CD, **1**.21
Ground connections, microprocessor, **6**.11

Helium-neon laser, **2**.18, **6**.49, **9**.46
HF (high-frequency) eye pattern, **5**.8

Interleaving, **1**.10, **2**.5, **2**.9
Interlocks, **4**.2
Isolation transformer, **4**.1

INDEX

Laser:
  adjustments, **8.6**
  radiation, **4.4**
  safety, **4.3**, **4.6**
  troubleshooting, **8.20**
Laservision, **2.18**
Leakage-current, checks, **3.1**
Lens, objective (*see* Objective lens)
Loading, **7.17**
  CD, **5.3**
  CDV, **6.1**
LP (long playing), **1.2**

Meters, distortion, **4.8**
Microprocessor:
  clock, **6.11**
  ground connections, **6.11**
  power connections, **6.11**
  reset connections, **6.11**
Microswitch adjustments, **7.13**, **7.25**
Moisture condensation, optics, **4.2**
Motor, replacement, **7.11**
Movie-film conversion, **2.20**
Multiplexing, **2.3**

Noise reduction, **6.40**
  CX, **1.22**, **6.46**

Objective lens, **1.4**, **2.15**, **4.2**, **4.5**, **4.10**, **5.5**, **5.7**, **6.16**, **7.5**, **7.14**
On-screen display (OSD), **6.40**
Optical pickup:
  CD, **1.7**, **2.12**, **4.5**, **4.10**, **5.5**, **5.7**, **7.5**, **7.14**
  CDV, **1.21**, **2.14**, **2.18**, **4.2**, **4.5**, **4.10**, **6.15**
OSD (on-screen display), **6.40**
Oversampling, **1.12**, **2.11**

PCM, **1.3**, **1.7**
Pickup adjustments, **8.8**
Pickup:
  optical (*see* Optical pickup)
  rotating arm, **1.5**
  slide (*see* Slide pickup)
Power connections, microprocessor, **6.11**
Preamp, CD, **1.10**
Pulse train, **1.11**

Quantization, **1.11**, **2.3**

Radial tracking, **2.20**

Radiation, laser, **4.4**
Receiver/monitor, **4.7**
Remote control, **3.14**
  CDV, **6.3**
Replacement:
  belt, **7.8**, **7.23**
  motor, **7.11**
Reset connections, microprocessor, **6.11**
Rotating-arm pickup, **1.5**

Safety, laser, **4.3**
Sampling, **1.11**, **2.1**
Shipping screw, **3.2**
Slide pickup:
  CD, **1.4**, **2.12**, **4.5**, **4.10**, **5.5**, **5.7**, **7.5**, **7.14**
  CDV, **1.21**, **2.14**, **2.18**, **4.2**, **4.5**, **4.10**, **6.15**
Special effects, **6.43**
  digital, **1.23**
  troubleshooting, **9.45**
Spectrum, CDV, **1.18**
Stereo system, standard (shop), **4.8**
S-VHS, **6.4**

Tangential tracking, **2.20**
TBC (time-base correction), **1.22**, **6.4**, **6.31**, **6.36**
TeD, **1.2**
Test discs, **4.7**
Tilt control, **6.26**
Time-base correction (*see* TBC)
Track:
  CD, **1.4**
  CDV, **1.21**
  jumping:
    CD, **5.19**
    CDV, **2.22**
Tracking, **1.9**, **2.12**, **2.17**, **5.11**, **5.19**, **6.21** to **6.26**
  adjustments, **8.8**
  radial, **2.20**
  tangential, **2.20**
Transit screw, **3.2**, **7.5**
Trouble symptoms versus adjustments, CDV, **9.35**
Troubleshooting:
  CD:
    autofocus, **8.18**
    checks, **3.11**, **8.4**
    laser, **8.14**, **8.20**
    mechanical, **8.11**

Troubleshooting, CD (*Cont.*):
   preliminary, **3.**9, **8.**4
   signal processing, **8.**16
   tracking, **8.**20
   turntable, **8.**22
CDV, **9.**1

Turntable:
   adjustments, **8.**9
   troubleshooting, **8.**22

Unloading, **5.**3, **7.**7
VHD (video high density), **1.**2

ABOUT THE AUTHOR

For over 39 years, **John D. Lenk** has been a self-employed consulting technical writer specializing in practical troubleshooting guides. A long-time writer of international best-sellers in the electronics field, he is the author of 72 books on electronics which together have sold more than 1 million copies and have been translated into eight languages. Mr. Lenk's guides regularly become classics in their fields and his most recent books include *Complete Guide to Modern VCR Troubleshooting and Repair, Digital Television, Lenk's Video Handbook*, and *Practical Solid-State Troubleshooting*, which sold over 100,000 copies.